OXFORD MEDICAL PUBLICATIONS

Childbirth for men

Childbirth for men

HERBERT BRANT

Professor of Clinical Obstetrics and Gynaecology
University College, London

Oxford New York Toronto Melbourne
OXFORD UNIVERSITY PRESS
1985

Oxford University Press, Walton Street, Oxford OX2 6DP

London New York Toronto
Delhi Bombay Calcutta Madras Karachi
Kuala Lumpur Singapore Hong Kong Tokyo
Nairobi Dar es Salaam Cape Town
Melbourne Auckland

and associated companies in
Beirut Berlin Ibadan Mexico City Nicosia

Oxford is a trade mark of Oxford University Press

British Library Cataloguing in Publication Data
Brant, Herbert
Childbirth for men.——(Oxford medical
publications)
1. Pregnancy   2. Childbirth
I. Title
618.2'00240431   RG525
ISBN 0–19–261450–9

Library of Congress Cataloging in Publication Data
Brant, Herbert A.
Childbirth for men.
(Oxford medical publications)
Includes index.
1. Childbirth.   2. Pregnancy.   4. Postnatal care.
I. Title.   II. Series.
RG25.B658   1985   618.4   84–20722
ISBN 0–19–261450–9 (pbk.)

Set by Wyvern Typesetting Ltd
Printed and bound in Great Britain by
Biddles Ltd, Guildford and King's Lynn

'There is nothing either good or bad but
thinking makes it so'

Hamlet

To my wife, Margaret

Contents

Acknowledgements

I wish to thank Oxford University Press for suggesting the book and directing its inception, and for their invaluable suggestions and editorial work.

I am grateful to the parents who so willingly agreed to be photographed with their babies, but with apologies to Rhys for being called Ben on the cover! I am also grateful to the many parents who over the years have contributed to my understanding of obstetrics.

I am also grateful to medical, midwifery, and physiotherapy colleagues at University College Hospital for suggestions and discussion.

Finally, I am especially grateful to Mrs P Cowburn, my secretary, for all her work in typing the manuscript.

Introduction

The importance of the involvement of the father as a partner in the childbirth process is now well accepted in most Western communities, but there remain widely differing attitudes to the father's role. Traditionally, men have tended to feel a little peripheral to pregnancy, childbirth, and child-rearing, and therefore somewhat excluded. There has been a tendency to look upon these as women's work and to regard too close an involvement as somewhat unmanly. However, we are now more aware of the advantages which may accrue to the family from the greater participation of the husband. After ensuring that both mother and baby emerge healthy and unscathed, the most important outcome of childbirth is the establishment of warm and strengthening emotional bonds between the parents and baby, and a consolidation of the emotional ties between the parents. These emotional relationships are the basis for the health and happiness of the family. My experience suggests that, although by no means essential to this process, close involvement of the man with the trials, tribulations, and joys of pregnancy and childbirth can be very helpful in ensuring the continuing strength of these ties.

Although having a baby is a common and natural experience, it is by no means always an easy one. There may be hidden dangers, difficulties, and pitfalls. For each person and couple, having and nurturing a baby is unique—no two experiences are the same and the spectrum is very wide. There are, however, generalizations and general principles which can act as guidelines. Although you can never be fully prepared for the future, you are put on the right track by learning the general principles and something of the possible spectrum or range of these future experiences. Even though the future may not be quite as expected or planned, these guidelines can be the basis from which you adapt and adjust to the realities of the experience you and your wife happen to have.

The aim of this book is to present childbirth more from the man's point of view. It will give you background information and some idea of how you may fit in: how events and situations may affect you; and what you can do to help your wife, your baby and yourself. I have for some thirty years been interested in this aspect of obstetrics—I know how much you can gain and how much your wife will appreciate your involvement in this fascinating phase of life.

In presenting childbirth and child care, there is a temptation to present the ideal rather than the reality, and to present the best picture rather than the whole spectrum of possible experience. This has the unfortunate effect of encouraging unreal expectations with consequent disappointment and disillusionment. Expectations are of prime importance. If these are more realistic, then you and your wife are much more likely to emerge unscathed emotionally and to have less difficulty in becoming reconciled to the particular course your wife's childbearing happens to take. My experience has confirmed the importance of realism rather than poetic idealism. I hope you find the book interesting, informative, and helpful, and of course I shall welcome comments, suggestions or advice which may improve future editions. Please write to me care of the Department of Obstetrics, University College Hospital, Huntley Street, London, WC1.

This book is written primarily for fathers expecting their first babies, although those who have already had one will find much of interest. I devote a separate chapter to second or later babies. In it, I highlight the ways in which second pregnancies, labours, and deliveries differ, and I also cover a few additional points which may interest you. In general everything is much better the second time around—indeed, if all women started childbearing with their second pregnancy there would be little call for this book!

I wish to mention two small problems which arise. The English language does not have a word meaning 'he or she'. Although it has been traditional to use 'he' when the sex is uncertain, there are those who consider this a reflection of unacceptable male dominance. To use she/he or s/he is clumsy for the reader, so I intend to use he or she at random in referring to the fetus or baby and leave it to whoever wishes to count and decide which appears more often!

Secondly, I am fully aware that not all fathers are married to the woman having the baby, but I am also aware that 'partner' does not adequately describe the status of husband or wife, so will favour the majority and use the terms husband or wife with advance apologies to other 'partners'.

Technical terms are explained when they first appear in the text and also in the Glossary on pages 199–201.

1 Having your baby as 'naturally' as possible

The aims of obstetricians and midwives are to encourage and facilitate natural processes when these are compatible with the safety of mother and baby. A large amount of the effort of these professionals is usually also aimed at promoting the emotional interests of individual women, their husbands, and babies. More importantly, professionals also aim to be kindly, caring, and good-humoured people.

Everyone knows that having a baby is not a new phenomenon, so why all the fuss? Well, I think it is helpful first of all to look a little more closely at what I am pleased to call the philosophy behind nature, and natural pheonomena. There is from time to time a resurgence of 'nature knows best' or a 'return to nature', as though what is natural is always good. There is an assumption that people should depend on natural processes and that these can be relied on to work in the interests of that individual.

I agree that all natural or biological processes work well most of the time. Seen as a whole they are admirable for the requirements of some grand design encompassing all species. It is as though nature throws plenty of chances into the ring in the knowledge that sufficient are bound to flourish. If you plant carrot seeds in your garden, most will germinate, but some will fail. Of those that start growing some wither easily, some form small deformed roots which are not worth eating, but many go on to form strong plants which can withstand the vagaries of the weather and, if allowed, will go on to flower, be pollinated, form seeds, and start the next generation.

Human reproduction is not dissimilar. When a couple are having intercourse regularly there are many millions of sperm chasing the egg that is liberated most months, yet, on average, conception takes several months. When an egg is fertilized it has only a 50 per cent chance of embedding in the lining of the uterus and forming an embryo. Of these pregnancies which start, 10–20 per cent end in miscarriage during the early months—almost always because of a chance error in the early development of the baby (the embryo or fetus) or of the afterbirth (placenta). After this time almost all pregnancies go on to produce a healthy, fully grown baby who does well and develops into a healthy adult. But there are some late miscarriages, some babies deliver prematurely, occasional babies are stillborn, and a few have a serious developmental defect. The placenta may not function adequately in late pregnancy or labour, and there are a whole host of occasional problems of one sort or another.

Nature works as though its interest is in the species as a whole, with the individual being of little account: as though it is the survival of the species—the survival of man—which is the prime concern. This is achieved magnificently but this attainment is of little comfort to the individual for whom the process has ended unhappily. In other words, nature has an inbuilt inefficiency and unreliability so that inevitably some individuals are going to be very disappointed in the results of their reproductive efforts. These inefficiences and unreliabilities, or the possibility that they may arise, are the adverse side of natural processes. If all doctors, midwives, and nurses died out tomorrow, the human race would continue. But the justification for having these professionals is to try, in the interests of individuals, to make up for some of the inefficiencies and defects of nature.

There are really no places where birth proceeds entirely 'naturally', as even what we may regard as primitive communities have their own systems and traditions built up over the centuries. There has probably never been a time when instinct ruled supreme such that having a baby was truly 'natural'. Man seems always to have had leanings towards organization, ritual, and ceremony and his intelligence has favoured the handing on of traditions, while his imagination has favoured experiments and development.

It is, however, worth noting what happens in some of the less-developed (by our own standards) communities. A few years ago I was talking to a Professor of Obstetrics in Ghana who was trying to improve obstetrics in rural areas of his country. In Ghana, 75 per cent of all deliveries are either unattended or attended by untrained personnel (relatives or traditional birth attendants who are usually male). In rural areas this percentage is much higher. For the country as a whole it is estimated that 5 women in each 1000 delivering die, and again the figure is higher in rural areas (this horrifying number compares with 1 in 10 000 in England and Wales). In Ghana, the perinatal mortality (stillbirths and deaths in the first week) is over 100 per 1000, and above 200 per 1000 in some rural areas, compared with 11 per 1000 in England and Wales. The Professor had nothing good to say about the true squatting position for delivery because of the serious injuries to the pelvic floor which he has found more common with this position. I am told by those who have practised there that childbirth gave similarly unsatisfactory mortalities and injuries in rural Sri Lanka: but results are improving rapidly with the acquisition of obstetric services. I recently visited a hospital in Kuwait, where they still have Bedouin women coming in from the desert. They come, among other things, for the management of serious injuries sustained during delivery in the true squatting position at home in the desert. The doctors in Kuwait were amazed to learn that some European women wish to return to this system. No doubt in our hospitals we can do better with careful control as the baby's head emerges, but this control is so much more difficult to effect when one cannot see clearly the perineum (the area between the vaginal opening and the anus) where these bad tears can occur. It is also not as easy to make the protective cut (episiotomy) when it is needed.

Recent history

You might be interested to learn a little of the history of some of the systems which have been enthusiastically introduced over the years to prepare women for labour and delivery. Their proponents have usually claimed: 'we now have the answer to making the experiences less painful and more positive and rewarding'. Almost all have had good points to make but have foundered by building up unrealistic expectations; by making exaggerated claims about achievements; by failing to explain the limitations of a system; and, most important of all, by failing to make clear distinctions between primigravid (first) and multigravid (second or later) labours. Many have dealt mostly with multigravid women who have, not surprisingly, had much better experiences than they had had with their first delivery. It can be quite misleading to attribute a better experience to the preparation system used the second time. True believers who initiate new systems are inclined to see and hear only what they wish to see and hear, and only that which fits conveniently with their theories.

Although we will continue to modify it, the antenatal education programme used currently in our hospital is a distillate of our own experience, the experiences of our patients, and the work of the many others who have contributed to our understanding. The contributions of 'orthodox' obstetric practice and research have been crucially important in improving safety for mother and baby and in improving techniques such as epidural anaesthesia and those used for induction of labour. We must have this happy mix of science, psychology, and humanity.

Towards the end of the last century, travellers and missionaries returning to Europe gave accounts of the ease, painlessness, and safety of labour and delivery among primitive peoples. These accounts had some influence. The degeneracy of civilization and luxurious living in Western societies were blamed for the sufferings of women in childbirth and for the relatively high maternal and perinatal death rates which prevailed in Europe at that time. (It has since been realized that these optimistic reports were quite inaccurate, and probably arose from seeing quick easy deliveries by occasional multigravid women.)

Women were advised to try to emulate the natural way of life and it is not therefore surprising that early attempts at educating women were through the vehicle of physical fitness and exercise. In 1914, Fairbairn, an obstetrician at St Thomas's Hospital London, asked Minnie Randall (the midwife principal of the School of Physiotherapy) to provide post-natal restorative bed exercises and later encouraged her to give pre-natal instruction. He hoped this would lead to normal function in labour and enable the patient to utilize her 'powers' and to have more confidence in them. This suggestion linked up with his opinion that mental states had an inibiting effect on the way the uterus worked in labour.

There has never been evidence that the majority of exercises which pregnant women have been urged to practise confer any physical benefit on labour or the pain of labour. Physical fitness is not a primary requirement for labour. In my experience, ballerinas, athletes, and gymnasts do not have easier labours or a smaller range of problems. The 'tailor position' exercise which is supposed to stretch the perineum cannot be effective. It involves the muscles and tendons of the inner upper thigh. It is not necessary to try to make pelvic joints and ligaments more flexible since they are already so influenced by the hormones of pregnancy. Even without 'flexibility' exercises ligaments often become *too* lax with resultant back pain, pelvic pain, and sciatica. These problems are occasionally quite crippling in late pregnancy. However, there is little doubt that women who remain physically fit during pregnancy will make a quicker recovery after delivery or caesarean section and will be better able to cope with baby care and any other stresses.

In 1933, the Englishman Grantley Dick-Read published his first book *Natural childbirth*. He had in many years of medical practice been dedicated to relieving the pain and dread which seemed to accompany human birth. He took up the concept of easier childbirth in primitive peoples and, swayed by the popular demand of the time, included exercises in his scheme of training. Unlike some of his disciples, he considered them of limited value and paid more attention to women as thinking, feeling individuals with their own physical and emotional problems. A struggle to increase the physical safety of obstetrics had quite rightly been the main concern in the mechanistic approach of most obstetricians up to that time. Dick-Read's work was a reaction to the dehumanization which then accompanied the management of labour.

For first-stage contractions, Dick-Read made use of the calming and distracting influence of relaxation and controlled breathing. He observed the powerful effect on women in labour of being attended by calm, sympathetic, and supportive staff. Although women were much relieved in this way, he looked for the real cause of the pain and distress of labour. He theorized that this was the fear and expectation of pain built up over the centuries. This led to his fear–tension–pain cycle postulate. He believed that fear could be combated by simple explanation, by the mother's learned relaxation, and by the quiet but steady understanding and support from those in contact with her during labour.

His interest was not only in minimizing pain but also in upholding the joy and dignity of motherhood. This found a ready response in many thousands of women. Unfortunately, like so many who have contributed since his time, he did not adequately distinguish between women having first babies (primagravidae) and those who have already borne a child (multigravidae). He was therefore falsely optimistic about his methods and furthermore underestimated the contribution of his own personality.

Labour and its accompanying pain did not prove as easy to conquer as Dick-Read had hoped but he did not reject, as some who have followed him, the support of science or the use of analgesia when necessary. He did urge the study of, and resort to, natural means as a first priority.

It should be remembered that Dick-Read worked without the advantages of pethidine or of epidural blocks. Prior to the introduction of pethidine in 1939 and epidurals a little later, patients were often confused by heavy sedation with morphine and hyoscine. When pethidine is given to a woman before she becomes too distressed by pain, and particularly if she has been prepared for labour, we find that her ability to co-operate improves. We can therefore have a more relaxed attitude to the use of this analgesia.

Another pioneer who rather later, after coming into contact with Dick-Read and Minnie Randall, became a strong influence was the physiotherapist Helen Heardman. In 1944, some years after the birth of her first child, she joined the School of Physiotherapy in Leeds and organized with the Maternity Hospital staff a 24-hour service of physiotherapists for the labour wards, and here she learnt about the realities of labour. She ensured, through the vigour of her personality, a rapid increase in the popularity of antenatal training classes. She joined the staff of University College Hospital, London, in 1948, as our first Obstetric Physiotherapist. (University College Hospital was, by the way, the first and I think the only medical school to invite Grantley Dick-Read to lecture to the students.)

Mrs Heardman understood how essential it was to reinforce antenatal teaching with continual support and encouragement during labour. She believed physical training to be the most important aspect of training for labour, but she again underestimated the contribution of her own personality and enthusiasm. She helped, without intending to do so, to establish a standard pattern of antenatal classes with sets of exercises. This tendency was exaggerated when the 'teacher' was a physiotherapist without extensive first-hand experience of being with and observing labouring women. Exercises were easy to put across and some teachers welcomed this security. Patients thought they were being taught something important and, in this setting, easily developed fairy-tale images of themselves in labour. When faced with the reality of labour, these images were often shattered, and the resulting disillusionment sometimes made them bitter and unhappy.

Just as British practice evolved, coloured by the people who helped develop it, and influenced by characteristic British attitudes and traditions, so the Russian psycho-prophylactic method evolved, chiefly through the influence of physiologists and neuropsychiatrists, from hypnosis and hypno-suggestion. Systematic influencing of the woman's mind with a view to preventing labour pain began in the Soviet Union in about 1920. Later, Pavlov's teachings about the conditioned reflex and his theory of human behaviour were applied to eliminate pain by conditioning training. Intensive education of the pregnant woman and all who came into contact with her, and indoctrination with the concept that 'there is no pain in labour' were the basis. This was combined with the development of positive ideas and emotions to eliminate pain perception. (Very effective—at least in theory!)

This work was further developed in France during the 1950s by Lamaze and then Vellay, who emphasized that 'there is no pain' and taught rather complicated theories of nervous system function. Strict attention to detail was essential. This gave a convenient let-out for the system if labour proved difficult or painful in a particular case. The first question to the patient might be 'Did you practise exactly as instructed?' As very few are that conscientious, failure could easily be attributed to the patient rather than the method. Women were either 'successes' or 'failures'; not exactly uplifting for those starting motherhood labelled as 'failures'.

In 1956, I was one of the very few doctors in New Zealand practising 'natural childbirth'. A returning New Zealander, who had her first baby in Paris in the care of Dr Lamze booked with me for her second delivery. I shall never forget her first remark, uttered with some emotion: 'I was one of Dr Lamaze's failures'. Her husband, who had also undergone the intensive 'there is no pain' preparation was present throughout this extremely painful labour. He was still, a year later, quite shattered by the experience. I hadn't then heard of Dr Lamaze and his methods, but you can imagine I didn't rush to embrace them! I claim no great credit for the complete success of her subsequent (twin) delivery, as this was her second and therefore could have been expected to be much easier.

Regimes used in France have since been modified and simplified so that much of this criticism no longer applies.

Because the French variety of psychoprophylaxis has been further modified in various ways in Britain, USA, and other countries, as well as in France, it has come to mean widely different things in different countries, cities, towns, and clinics throughout the world. This is especially so in the details of complicated breathing regimes which most primigravidae are not able to carry out when labour becomes difficult.

All varieties of preparation are aiming to further the interests of women and their husbands in pregnancy and labour, and in a wider sense to improve the quality of family and individual life. Undoubtedly the qualities of the teachers are more important than the methods used. It is essential not to put undue pressure on women to 'perform'. Women and their husbands are best prepared where they will be delivered and by those familiar with the local philosophy for the management of labour and delivery. In my opinion the worst and most unkind approach is to try to influence the general approach to the management of labour in a hospital or clinic, while a woman is pregnant or in labour. This puts excessive pressure on her, is likely to destroy the trust she must have in her professional advisers, and can alienate the staff from her. Not surprisingly, with this destruction of trust and confidence she often ends up needing all the interventions she had so wished to avoid. It is important to have pressure groups as they can have important influences on practice but in my opinion these influences should not be manifested *via the patient* while she is pregnant or during childbirth. She is too vulnerable at these times.

Recent developments

More recently, Leboyer of France introduced the 'birth without violence' concept. His principal contribution was in making people more aware of the importance of treating the newborn baby gently. There were, and probably still are, places where this is inadequately appreciated. Unfortunately, he also had some unhelpful suggestions: according to his philosophy, during the birth the baby undergoes a traumatic experience and should therefore be put in a bath of warm water where the obstetrician should massage her for half an hour to simulate uterine contractions. The idea was that the feeling of uterine contractions and of being in water were familiar to the baby and would therefore lessen the shock of transition. I suggest that when the baby is inside she is meant to be in water, and when outside she is meant to be out of water. Is it not much more appropriate that the mother or father cuddles and comforts the baby?

Delivering the baby on to the abdomen is fine if that is what the mother wants. When he is in this position, to allow the cord to remain untied as it continues to pulsate can drain the baby's blood into the placenta and run a risk of him becoming anaemic. (This fact had been established many years previously.) No doubt some of Laboyer's ideas have since been modified.

The term 'active birth' has been promoted as though its concepts are all new. It has been well known since the mid-1960s that lying flat on the back during labour is undesirable for several reasons, none of which apply to lying on the side or to a semi-sitting position. There has been good evidence (though not universally confirmed) that labour progresses more quickly in the semi-reclining or erect positions than when lying flat on the back, but there is no study to suggest that lying on the side is unfavourable. Women often like to walk about or sit in a chair in the earlier part of labour. No one position is preferred for any length of time and it is the ability to move about and to change postures which seems to be helpful. However, there is no reason why a woman should be urged to be up and about when her desire is to lie on her side when labour is more advanced.

There is a widely misquoted X-ray study, undertaking in 1958, of the area of the outlet of the pelvis. This shows that the area between the bones at the outlet of the pelvis is increased by up to 30 per cent in the sitting position, when compared with the same area when the woman lies on her back. The study had nothing to do with the true squatting position. The claim that this X-ray study showed the area of the pelvic outlet to be greater in the squatting than the semi-sitting position is not true. The squatting position has the distinct disadvantage of not allowing observation of the perineum where tears are likely to occur as the baby emerges.

Another variation is being suggested by the Frenchman, Odent. He has been very successful in relaxing the environment of his hospital and claims to encourage 'instinctual' behaviour. However, he seems to espouse a certain rigidity with his 'preferred delivery position' and his earlier insistence that no pain-relieving drugs may be given during labour in his hospital. It is difficult to see what is 'instinctual' or natural in a land animal (man) delivering a baby into a bath of contaminated water! It is interesting to note that aquatic mammals, such as the seal, which have a choice, always choose dry land for giving birth.

I note with considerable anxiety that yet another Guru has arisen in Russia. Igor Tjarkovsky claims great virtues for delivery under water. Babies are then trained in the first three months much as the mother porpoise might train her offspring. Despite claiming that the baby can safely stay under water for ten to fifteen minutes after birth, babies are said at three months to have the same ability level as a normal one-year-old, and they later grow into strong independent children!

I am reminded of Dr Heynes of South Africa and his decompression suit, a craze which reigned in the late 1950s and early 1960s. Dr Heynes had women in a decompression suit for varying periods of time repeatedly during pregnancy. This was said to improve the blood supply to the placenta and the babies were said to be of above average intelligence and above average development at two years. The suit not only improved backache in pregnancy but was of great value for pain relief in labour, etc. We were plagued by people clamouring for this new miracle of technology. Businesses sprang up in Knightsbridge to hire out machines. We had several in our hospital. When properly controlled trials were set up, results showed no benefit whatsoever. It was the same old story of motivated women volunteering quite reasonably in the hope of benefitting their labour and offspring. They then gave their 'special' babies extra attention and stimulation. But similarly motivated and similar-class women had equally good results without decompression. The benefit in labour was due to extra attention and kindly care rather than decompression. The need for pain relief was not reduced in controlled trials.

Conclusions

It is difficult to steer a middle course in accepting what is helpful and rejecting what is not. One of the real problems of such regimes which claim to have 'the answer' to obstetrics is that the regime sometimes becomes important to the exclusion of the individual. Often it is the character and enthusiasm of the people involved which is much more important than the regime. Character and enthusiasm are not so easily transplanted. It can be a mistake to encourage your wife to invest her energies and enthusiasm in a system which is not practised where she will actually be having her baby.

It is because it is unlikely that we will develop the ultimate solution to the problems involved in having a baby that enthusiasts and gurus will continue to arise with their heartfelt convictions. Most new regimes or systems come to attention via a book written by the enthusiast, in magazine or newspaper articles, or on television. All tend to present highly appealing, one-sided, laudatory, and 'pro' accounts. The true credentials of the guru are not mentioned and there has usually been no serious attempt to assess the true worth of 'the system'. Associated disadvantages, defects, and disasters are glossed over or not considered. It would seem that a sober appraisal would reduce media appeal, the sales of the magazine, or the audience ratings. Newness and novelty appear to be all-important. Perhaps one could be charitable and say that the principal problem is usually that the reporters cannot be sufficiently well versed in the subject. I think having a baby is too serious a matter for this trivialization, but I can understand why you may say to each other: 'That looks lovely', 'Why can't we have our baby like that?', or 'Why don't doctors let you do it like that in our hospital?' Our reservations concern the safety of mother and baby, and doubts about exaggerated claims. Birth is usually a wonderful and moving experience—at least for those watching, and often for the mother—but this may have nothing to do with a particular position or method being advocated. Meeting the needs of individual women with dedicated and caring staff is the essential ingredient for good results.

Many women currently pregnant tell us that they feel very pressurized from all directions to 'do it naturally'. You husbands may see a film in which a woman reports that labour was not painful. You can easily compound this pressure by saying, 'See, she said it isn't painful'. Resist this temptation and keep an open mind. It is very rare for a *first* labour to be painless and in almost all *first* labours women *need* and *have* pain relief with pethidine, an epidural, or gas and oxygen for the end of first stage.

2 Second (and later) babies

I have placed this chapter near the beginning of this book so that those of you whose wives are having a second or subsequent baby (such women are called 'multigravidae') can interpret the following chapters for your own situation. As mentioned in the Introduction, I have written the book, with the exception of this chapter, more for men whose wives are having a first baby. However, I am sure, despite this, you will find many of the chapters interesting.

Experiences of the first pregnancy and labour obviously colour expectations the second time around. There may therefore be less or perhaps more anxiety. You and your wife should seek to have any unexplained or unresolved problems of the first labour and delivery ventilated through discussion with your doctors or midwives. This can be an important function of antenatal classes, especially if they are held for multigravidae separately from those of first-timers. The second time around the uterus works more efficiently and, as tissues have previously been stretched, labour is usually shorter and there is usually less need for such things as drips, pain-relieving substances and forceps or caesarean delivery.

The woman who has already had a baby is in quite a different category from someone embarking upon a first pregnancy. If she has a tendency to any problem such as raised blood pressure or having small babies, then the problem is declared and, even though it may not return, one is alerted to this possibility. The weight of a second baby averages 100g more than the first, but even if there is no definite pattern a second baby usually delivers more easily irrespective of its weight.

Everything tends to be more efficient and favourable the second time around. However, despite this, the need of your wife for comfort, reassurance, and appreciation from you, her husband, is still there. The first pregnancy seems to attract all the concern, sympathy, expectation, and excitement and often there seems none left over for the re-run. To your wife this is a new and different experience, even though she may look and seem much the same as in the first pregnancy. To make the little extra effort at this time will be very much appreciated and is so important in cementing and developing your relationship. I find it very easy, when walking along a post-natal ward, to know who has had a second or later baby. Whereas those with first babies are peeping out from behind what looks like a florist shop, those with a second may have a couple of faded petunias or a dried-up cactus!

How does the pattern of labour differ from a first labour?

The pattern of labour is usually quite different and in most ways more favourable. You cannot, by the way, have failed to notice my frequent use of words such as 'usually', 'often', 'most times' etc. There is nothing absolute about any natural phenomenon. All one can hope to do in a book of this size is to follow the main stream and mention some of the exceptions. Variations are legion and can be appreciated only after long experience. I want to inform you as much as is practicable, but I cannot give you a medical training and then ten years full time in the specialty! Reading this book and attending a few classes won't make you an obstetrician. This is the justification for having our advice and guidance and we, in our turn, have to look upon each person as an individual with her own variations on the obstetric theme. We gather information from the history of the present situation, from the past history, and by current examination. We then make predictions which, however, must be constantly updated as events unfold. When I suggest that second labours are more favourable, events may prove me quite wrong in an individual case. It is better to expect the most likely but to maintain a certain flexibility of expectation.

Labour generally tends to be rather shorter, and the more difficult, painful part of labour is very much reduced. However, the sudden change in the rhythm of labour, the powerful contractions, and rapid progress which may characterize late labour can start suddenly and can be frightening and overwhelming even if anticipated. A woman may feel as if she is being buffeted about by strong contractions over which she has no control. Sometimes it is this lack of ability to control the situation which may for some be more difficult with a second labour. Even though the first was usually longer and overall more painful, it probably developed with a slower and sometimes more controllable rhythm. Contractions with the second tend to seem relatively mild for most of the time but can suddenly become very strong.

The phase of strong contractions may last only ten to twenty minutes but may be two to three hours even when all is going well. The second stage may be only one or a few strong prolonged contractions, but very occasionally it can be much more protracted and end with a forceps or even caesarean delivery. Sometimes the pushing doesn't seem to be going well and your wife may be holding back and pushing into her throat instead of downwards with the feeling of letting go and opening out. She may need firm instructions to give her the necessary confidence. Although the stretching at the opening is often less painful, if delivery is more rapid or if there have previously been stitches, it may be more painful. It is usually better and safer when there is significant scarring for her to have an episiotomy with a local anaesthetic.

If you want to be present at the birth, don't leave the labour ward once your wife is settling into the rhythm of labour. Things can change quickly and there may not be time to get you. Once your wife is in hospital, decline a midwife's suggestion that, because your wife will be ages, you might as well go home and have a sleep.

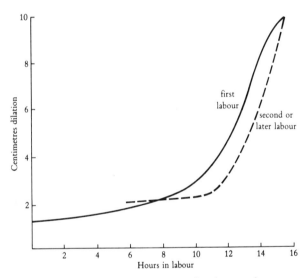

Second labours are shorter but finish strongly

Encourage your wife to be very optimistic about the course of her second labour, because almost all women have better labours and have normal deliveries the second time around, provided there isn't a known reason for having a repeat caesarean section.

It is not easy for you and your wife to know when to come into hospital in good time for care in labour and for delivery, since the course of labour is more unpredictable. It is better to be in early than too late. We advise the woman to come in when she thinks she is in labour as she is usually right—this is usually when contractions are coming at intervals averaging 10–15 minutes for one to three hours. Follow the advice of your own doctor or midwife, and telephone the delivery staff for advice if you are uncertain whether or not to come in. Don't wait for strong contractions. Much better to have one or two false alarms than to deliver on the way! Women having second babies are sometimes advised to stay at home until contractions are strong or coming at five-minute intervals. I consider this bad advice as no-one can predict when the strong end of first-stage contractions will start or how quickly they will lead on to delivery.

Pain relief

For multigravid patients analgesic requirements must be thought out beforehand, since analgesics should be given more for pain which is anticipated rather than for pain which has become distressing. To wait until the pain is bad before asking for a dose of pethidine is leaving it too late, as it won't act in time to help with the difficult end of first stage (unless a doctor is on the spot and gives a small dose into a vein). You and your wife have to decide beforehand whether or not she will want pethidine and, if she does, then ask for it when she is sure she is in labour and will not be going home. It is then likely to be working effectively in time for painful contractions. The same applies to an epidural—it has to be inserted early in labour—one of the problems is that, even with the best of intentions, there may not be time to get it working in time to be helpful for a quick second labour.

The alternative is to opt for no analgesia, in which case your wife can fall back on gas and oxygen inhalations if she finds at the time she wants pain relief. In general, I feel that most people who *do* want to have some analgesia should opt for one injection of pethidine. Provided it is given early, as suggested, it is effective and has minimal likelihood of side-effects. You can get some idea of likely requirements for a second labour from the pattern of your wife's first labour—if it has been relatively quick and not too painful without analgesia or with only one or two doses of pethidine, then gas and oxygen will probably be sufficient.

Quite a number of multigravid women decide to use gas and oxygen only, should analgesia be required. Don't push this idea but if your wife is one of these she will need plenty of reassurance and encouragement. If the difficult part of her labour does prove unacceptable long and painful then she can still have pethidine or an epidural. However, there is often difficulty in getting them working effectively in time to be useful for pain relief. In deciding whether or not to have them at this late stage, be guided by the midwife or doctor, who will examine vaginally to help decide how much longer the painful stage is likely to last.

Previous caesareans

Even if your wife has had a long and dreary labour the first time, or has had a forceps or caesarean section delivery, she can have a relatively quick second labour. For those who have had a caesarean section the first time, a decision has to be made by the doctor whether to advise a repeat caesarean section, or to expect vaginal delivery this time. The decision depends on the reason for the first caesarean: whether there were any problems with or following it; the size of the current fetus; and the size and shape of the different bones of your wife's pelvis. He may need to take an X-ray to help formulate the decision. If a first labour has been, overall, less than say three hours, then I feel there is a case for inducing labour just before the due date for the second delivery if everything is favourable. This is just to make sure that delivery is in the hospital, if the plan is for a hospital delivery.

Breast-feeding

Your wife can be optimistic about her ability to breast-feed the second baby even if her attempt with the first was not successful. Like everything else, breast-feeding can be so much easier the second time. The milk supply establishes more easily, the nipples are softer, more stretchy, and easier for the baby, and your wife is now experienced and will be much more confident in handling this baby.

Antenatal classes

Antenatal classes for multigravid patients are better organized separately from those for women having first babies. Multigravid women have been over most of the material, have experience of labour and child care, and haven't the spare time to repeat a whole series of classes. We arrange two revision classes for them, concentrating on coping with multigravid labour, but women may attend any other classes such as analgesia or baby feeding, if they wish. A principal need in these classes is for women to be encouraged to have a free-ranging discussion about previous experiences, so they can have a chance to come to terms with these experiences if this hasn't already been done. This clears the mind and enables it to take in the new information about multigravid labour as set out above. Breathing, relaxation, and pushing techniques are revised.

   With the passage of time, the impressions of past experiences tend to change. Unfavourable experiences to which the woman has become reconciled tend to fade. Those with which she has not come to terms can plague her and become recalled as exaggeratedly worse. Good experiences sometimes tend to take on a rosy glow to a quite unwarranted degree. Tendencies to self-justification and self-glorification are quite reasonable developments but can further distort memories. This understandable lack of objectivity makes those whose experiences of childbirth are less immediate relatively unsuitable for explaining labour to those looking forward to their first.

Your first child

For a first child, the mother's second pregnancy provides an excellent opportunity for learning about human reproduction. All little children, whether male or female, have a baby in their tummy when their mother is expecting. They are not always pleased that there is going to be a second baby, as they already sense attention and concern being diverted to the one inside. Women often report getting the odd boot in the abdomen from the toddler as well as getting stroked or kissed. However, whether it is appropriate to have the child present when labour is distressing, and for the delivery, is quite another matter. Although occasionally the childs presence at delivery is advocated, I have grave doubts about its desirability, especially because what can seem initially like smooth progress can unexpectedly and rapidly deteriorate. The opinions of most psychiatrists are very much against small children being exposed to what they may interpret as the stressful and violent experience of their mother in labour and at delivery, even apart from the likelihood of blood and trauma. The same applies to older children when their mother is involved.

However, once the situation has been tidied up after delivery, the sooner all the family can be involved the better, if this is what the couple want. Many prefer to have a time alone with their new baby, welcome the opportunity to carry out a good inspection, if this hasn't already been done in their presence, and suckle the baby if this is their wish.

If a woman is hospitalized during a second pregnancy then it can be a real problem for the mother and child to maintain their good relationship if the home is some distance from the hospital. We encourage as much continuing contact as possible and the same applies when there is a hospital stay after the delivery. The younger the child, the more important this continual contact seems to be. It is generally much more emotionally healthy that both mother and child are upset at the end of visiting times than that the child loses this contact. In the latter circumstances, a child may seem docile and easily managed, but in the long run will sometimes suffer profound emotional disturbance.

Coming home

You will be surprised how much fitter your wife seems after the second baby: she recovers more quickly and is much sooner like her old self again. She seems a different person. She is much more confident and less anxious with the baby. There is much less hassle in getting feeding established and she will even be able to hold a conversation with you during the process! She will be able to carry the baby in one arm while she does other things. The stitches of an episiotomy or tear will worry her less and even a repeat caesarean section seems to cause less pain and upset. She may well say to you, 'Why couldn't my first labour and delivery have been like this?' and, 'Why wasn't I like this after my first baby?' The simple answer is because that was her first and this is her second. As I have said before, if everyone started with the second baby much of the difficulty of childbirth would disappear.

3 *Some background information*

You need to know something about biological structure and function if you are going to be rewardingly involved in the birth of your new baby. I shall therefore go over some aspects of pregnancy and labour, presenting them in a way which I hope you will find helpful.

The uterus

The uterus (or womb) is the focus of interest in pregnancy and during labour. It has two main functions. The first is to provide a safe, protected place in which the baby can develop and grow until she is mature enough to live separately. The second is to open at the appropriate time and to assist the mother to expel the baby—a process in which the baby probably plays no significantly active role.

The uterus is shaped like a hollow inverted pear with the opening below, where the stalk of the pear would normally be attached. The lowest part is constricted to form the cervix (or neck) of the uterus, and has a central canal little wider than the stalk of the pear. This canal is closed during pregnancy by a plug of thick jelly-like material—the mucus plug. This mucus helps in preventing bacteria (germs) from reaching the baby. The cervix is required to stay shut during the pregnancy despite the fact that the muscle of the rest of the womb is always active. It is therefore formed mostly of tough, relatively unyielding tissue. However, at the end of pregnancy, it must have softened and changed its quality, so that it yields to the stronger contractions of labour. It stretches and opens (dilates) during the first stage of labour, so that the baby can pass through the lower vagina in the second stage of labour to emerge at delivery. Towards the end of pregnancy and during labour, the lower, more fibrous part of the uterus thins and increases in depth, as the upper part thickens and shortens. This lower part is called the lower uterine segment.

The larger (upper) part of the uterus is essentially made of muscle which grows, thickens, yields, and stretches to accommodate the growing baby during pregnancy. At the end of pregnancy this upper part is very strong and becomes the principal source of power to deliver the baby. As a muscle works, it decreases its length (contracts). The muscle of the uterus is what we call an 'involuntary muscle', which means that it can contract without waiting for or needing orders from us. In fact, we cannot influence involuntary muscles or tell them what to do—they are not under our voluntary control or will. Another example of an involuntary muscle is that forming the wall of the intestine. You have about eight metres of intestine contracting away right now and doing its own thing—you become aware of this intestinal action only if contractions are unusually strong, such as when you have gastro-enteritis (or stomach 'flu) or have a lot of wind distending the intestine, or perhaps because you have taken too much bran! The uterus contracts

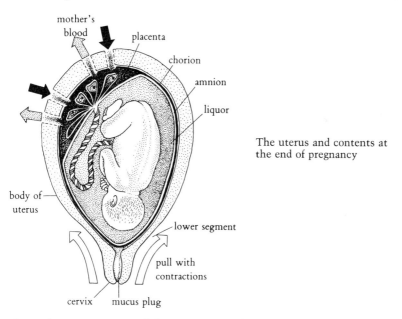

mother's blood

placenta

chorion

amnion

liquor

body of uterus

lower segment

pull with contractions

cervix   mucus plug

The uterus and contents at the end of pregnancy

throughout a woman's life, starting well before she is born. She is aware of these contractions only when they are strong. This occurs with menstruation when they may in a proportion of women give rise to very severe pain (called dysmenorrhoea). The woman also starts to become aware of uterine contractions in the second half of pregnancy as a tight feeling—sometimes an ache—in the lower abdomen or lower back. This is the uterus getting in a bit of practice for labour. If you put your hand on your wife's tummy at this time you will feel the womb becoming quite hard and pushing forwards or standing out more. How soon these contractions become noticeable and how strong they feel to your wife depends on her sensitivity to this sensation and on how strong the contractions are. Many women do not realize that these *are* contractions. They may feel the tummy go as hard as a board and the think the baby has got stuck in a funny position.

As stated above, when a muscle works (contracts) it decreases its length—an example is the biceps muscle, which draws the bones of the lower arm closer to the shoulder. When it stops working (relaxes) it returns to its original length. This is called a voluntary muscle because you can voluntarily control it and tell it what to do. The involuntary muscle of the uterus can be regarded as a series of strips of muscle attached to the cervix and passing up over the top and down the other side back to the cervix. These strips run in all directions. When they work, they shorten their length and, as the contents of the uterus are either liquid or solid (baby, placenta, and waters) and therefore cannot be compressed, the only effect of this contraction is to pull the cervix open. This action on the cervix is similar to your action in pulling a tight polo-necked sweater over your head, except that you would need to be upside down, as the cervix is the lower part of the uterus. The womb muscle works intermittently but, unlike the voluntary biceps muscle, it does not return quite fully to its original length when it relaxes. It undergoes a change to take up the slack a little each time, a bit like those winning a tug-of-war competition.

The contractions of late pregnancy are very similar to those of labour. Labour is really just the end of pregnancy when the uterine contractions become stronger, longer and closer together, and therefore more effective in pulling the cervix open. For some women the change from late pregnancy contractions to those of labour is fairly abrupt so that there is a definite time of onset of labour. With other women, one merges imperceptibly into the other so that there is no certainty about the time of onset of labour. All we can do is to make an arbitrary decision about the time of onset, so that we can put something in the space for the length of labour to keep the records in good order! In some women, the pregnancy contractions bring about much of the early opening of the cervix so that labour contractions have less to do and therefore labour is shorter. In others, all the work is done in labour which is therefore longer.

The placenta (afterbirth)

This is the organ of exchange between the mother's and baby's bloods and is therefore vital for the baby's continuing welfare during pregnancy and labour. It is circular in outline and attached to the upper part (or body) of the uterus. It has the shape of a flattened cake (plakous= placenta=cake). It is derived from the fertilized ovum and by the end of pregnancy is about 20 cm in diameter and 2–5 cm thick at the centre. It is a sponge-like structure which is filled by the mother's blood pumped in by about 130 arteries and leaving by as many veins after slowly oozing through with a flow of 500 ml per minute. The volume of mother's blood in the sponge at any one time is 250 ml.

The baby's blood is pumped by his heart along two arteries in the umbilical cord and passes through a series of minutely divided closed loops on the surface of the framework of the sponge, before returning to the baby's heart via a large vein in the cord. The baby's blood in these very fine loops is separated by a microscopically thin partition (membrane) from the mother's blood which bathes these loops. The two streams of blood don't mix, but because of the convolutions the total area of exchange has been estimated as 11 square metres. Oxygen, glucose, vitamins, and the other food substances pass from the mother's

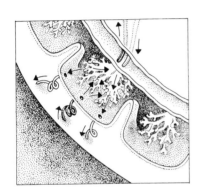 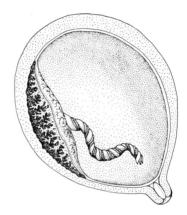

The network containing baby's blood is bathed by mother's blood in the placenta

to the baby's blood while carbon dioxide and the various waste products pass in the opposite direction.

The capacity of the placenta to supply oxygen and nutrients from the mother's blood tends to diminish progressively in the later weeks of pregnancy, even when the pregnancy is apparently normal. However, despite the growing needs of the baby, the reserve capacity of the placenta is usually such that it is more than equal to the task, even when the pregnancy continues two or three weeks beyond the expected date. However, in some women, the capacity can fail to continue to meet demand, so that the baby's growth pattern falls below normal and, in rare cases, the baby can die in the uterus before birth.

During each labour contraction, there is a temporary interruption of blood flow into the placenta because of the pressure built up in the uterine wall through which blood has to flow to get to the placenta. This is usually of little importance, but if it so happens that the capacity of the placenta to supply the baby's needs was just adequate before labour, it will become insufficient during labour. This can occasionally result in the baby dying during labour or being born brain-damaged because insufficient amounts of oxygen have been crossing the placenta from the mother's to the baby's blood.

This potential for placental failure, which can be present in any pregnancy, is one of the inefficiencies inherent in natural processes. This possibility is one of the principal reasons why skilled observation is needed in late pregnancy and in labour. Detecting early evidence of developing placental failure during labour can allow appropriate evasive action to be taken to reduce the chances of mishap. This may take the form of an episiotomy, forceps delivery, or caesarean section. Sometimes this evidence is not present and a baby can, without warning, be born in a depressed state because of the earlier lack of oxygen. He then needs expert management to give adequate oxygen while waiting for his own breathing mechanism to get going and take over.

This tendency to placental failure is more likely when there is an abnormal situation such as raised blood pressure, separation of the placenta, or advanced maternal age. These factors all tend to accelerate decline in placental capacity.

The membranes

Within the uterus, the fetus lies in a pool of water (the liquor) which is contained within two layers of thin but tough membrane with the consistency of a blown-up balloon. The outer membrane (chorion) is attached to the edge of the placenta, while the inner amnion extends to cover the fetal surface of the placenta right up to the root of the umbilical cord. The outer chorion away from the placenta is lightly adherent to the underlying uterine wall. These membrances usually break during the course of labour, towards the end of the first stage. They may, however, be defective and break before labour or even weeks before the baby is due. Very rarely, the baby can be born still within the membranes and the attached placenta (born in a caul). In the old days of sailing ships, a person who had been born in a caul was very welcome aboard for there was a legend that such a person could never drown at sea. This seemed to reduce the chances of the ship sinking!

These membranes are easily broken artificially when there is a need to do so and the cervix has started to open (dilate).

The liquor (waters)

This fluid, which is contained within the membranes, is also called the amniotic fluid or the bag of waters. It is within the amniotic cavity lined by the inner membrane—the amnion. The fluid which is formed initially by the amnion is supplemented after three months by the fetal urine formed by the developing fetal kidneys. The fetus swallows the fluid which is partly re-absorbed through the fetal intestine into the fetal blood stream.

Throughout most of pregnancy the liquor volume is several times that of the fetus, although during the seventh, eighth, and ninth months the volume relative to the fetus is progressively and markedly decreased.

The fetus has a specific gravity very little greater than the liquor, so that in it he floats freely in an almost weightless state (rather like astronauts in a space capsule). He therefore needs very little energy to practise all the functions which he will be called on to perform when he is outside. His at first feeble muscles start to move his upper limbs from five weeks after conception (seven weeks of pregnancy) and movements become progressively stronger and more co-ordinated from then on. By 12 weeks, he has fingers and toes, can actively kick his legs, curl and fan his toes, make a fist, suck his thumb, squint, frown, and open his mouth. His mother notices movements between 18 and 20 weeks when his kick is strong enough to be felt through the uterus on the inner lining of the front part of her abdominal wall (the peritoneum). As women having a second or later baby have felt movements before, they are more sensitive and know what to expect and therefore notice the movements (the quickening) usually three or four, but up to six or seven weeks earlier.

The liquor protects the floating fetus from external bumps and bangs because fluid disperses forces equally in all directions. If a pregnant woman has a considerable blow on her abdomen, the fetus is not affected. Imagine a leaf floating in a balloon filled with water. When you give the balloon a good bump the leaf just wafts gently up and down, because the water has acted as a shock-absorber. The fetus is protected in the same way. In late pregnancy your wife is liable to get into the habit of falling down the stairs as she trips easily. This is partly because she can't see her feet and yet may still wear high-heeled shoes, and partly because she has an altered centre of gravity. Her ability to recover if she does trip is impaired. Even if she breaks the odd limb or two she can be reassured that the baby won't come to any harm. It is easy for you to imagine the baby being fractured like a piece of china, but I have never known a baby to be damaged in this way. One woman told me she tripped on the gutter and fell on the way home from the fathers' class.

She wasn't too pleased when her husband said, 'Come on, get up, you won't have hurt the baby—the professor said so'. I don't mean to imply your wife shouldn't receive any sympathy on such an occasion and she certainly won't take that view!

While floating in the liquor, the fetus is subjected to equal pressure from all directions—both internally and externally. At first, its organs, such as the developing head, are very delicate but this equalization of pressure allows symmetrical development in a weightless environment. This would be impossible without the liquor. The liquor also ensures an even temperature the same as that of the centre of his mother's body. Because there is fairly free movement of fluids and chemicals from the fetus to the liquor, a sample of liquor can be used as a guide to the state of the fetus in rhesus haemolytic disease and many rare inherited disorders of the fetus.

The umbilical cord

This is the baby's lifeline. Blood containing waste products is pumped by his heart along the two arteries in the cord out to the placenta. Here waste is transferred to his mother's blood to be eliminated through her lungs and kidneys. Oxygen and food materials are collected and carried back to the baby in the large umbilical vein. At term (the expected date of delivery) the cord is normally about the same length as the fetus (54 cm), but can vary between 18 and 122 cm. It is fashioned in the form of a spiral (a bit like a dressing-gown cord), presumably because of unequal growth of the vessels, but possibly as a result of the twisting motion of the fetus.

As the blood flows through under high pressure, the cord is normally in the form of turgid undulations, so that any significant entanglement with the fetus is quite rare. About 20 per cent of babies have the cord loosely around the neck at delivery, but this does not often cause any real problem. The umbilical cord is long enough to allow the baby to move and swim about with considerable freedom.

I have been interested to learn that women sometimes think there is a connection between their own belly button and the baby. It changes shape in pregnancy—it flattens and may stick out and there is often a tense or pulling feeling there as the abdomen expands. This belly button is merely the spot to which your own umbilical cord was attached during fetal life. It is the scar left when the no longer useful umbilical cord shrivels and drops off after birth. Inside there is a gap between it and the underlying uterus and there is no connection.

The fetus and her development

For the first ten weeks after fertilization or the first 12 weeks of pregnancy, the developing baby is called an embryo. It is during this time that the various organs are formed and the early semblance of a baby is attained. After 12 weeks, the embryo becomes the fetus until after delivery. During fetal life, the already developed organs and body parts grow at different rates, so that there are alterations in proportions and appearance. A few new and minor organs such as hair, eyebrows, and fingernails form later than the main organs.

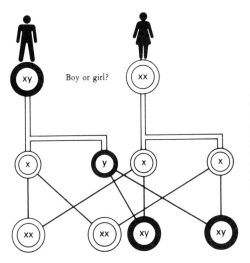

Men produce sperm with either X or Y sex chromosomes, while the eggs of women contain only X chromosomes. Fertilization with a Y chromosome sperm produces a boy, and an X chromosome sperm a girl

0–4 weeks of pregnancy

The ovum or egg grows in the ovary for the first two weeks and is then released and wafted in a stream of fluid into the fallopian tube where one sperm succeeds in penetrating and fertilizing. The single cell, which results from fertilization, is just visible to the naked eye and now starts the long process of multiplication by division as it drifts towards the uterus and burrows into the lining as a small ball of cells one week after ovulation. The outer cells begin to form the future placenta and membranes and the inner ones the embryo.

5–8 weeks

The organs and general shape of the embryo are forming rapidly. At the end of this time, the embryo is recognizable as a developing human and would fit inside a small plum. Her heart has been beating for two weeks, and these beats can be demonstrated on an ultrasound scan. There is a soft skeleton made of cartilage. The arm buds are sprouting and they already move feebly. She is floating in a small pool of liquor within the bag of membranes, and is attached to the placenta by the umbilical cord even though she's smaller than Tom Thumb.

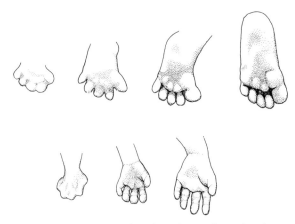

The upper limbs form from the fourth to the ninth weeks after conception
The lower limbs are a few days later

9–12 weeks

At nine weeks, your baby is forming her first bone within the cartilage framework. By 12 weeks, all the principal organs have taken on their final form and are functioning in at least a rudimentary fashion. From now on, they mature and practise their work in readiness for separate living. By 12 weeks, she weighs about 30g and measures about 8 cm. She has ears, nose, mouth, fingers, and toes. Her head is as large as all of the rest of her body put together. She swims around freely, but may lie quietly sucking her thumb. Her hands and feet have taken on their final form.

13–16 weeks

Your daughter now weighs 180g and is 20 cm tall. Her organs are now working well. Her kidneys are making urine which she passes from time to time to contribute to the liquor volume, while her intestine absorbs the liquor which she frequently swallows. We are not sure about her sense of taste! Breathing movements are present, so that the muscles of the chest and diaphragm are already starting to strengthen in readiness for air-breathing after birth. Her eyebrows and eyelashes have started to

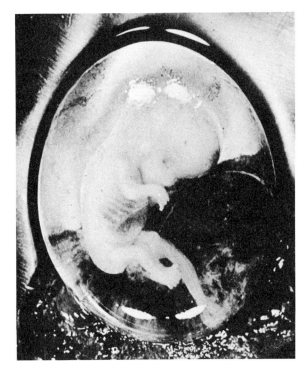

By 12 weeks the final form is achieved, although modification continues

grow: there is a bridge to her nose and her first set of teeth have started to form in her gums. Her finger and toe-nails have started to grow and her limb movements are strong and frequent.

17–24 weeks

The fetus now reaches 30 cm in length and a little over 450g in weight. She has hair on her head, her teeth are hardening, and her nails growing. Her eyes are opening. If she should be born at 24 weeks, it is just possible that she could survive and grow up normally if the best paediatric care were available.

25–28 weeks

Your baby (still officially a fetus) is 35 cm long and weighs 900g. She has fingerprints and is covered with fine down-like hair. She has developed a grip strong enough to support her weight if suspended in air. Although her heart movements can be made audible with an ultrasound device from about ten weeks, they are now easily audible through an ordinary doctor's stethoscope. She moves a lot and you may feel the kicks on your wife's abdomen, but she also has quiet periods, which don't necessarily correspond with your wife's sleep periods! Her skin is prevented from shrivelling in the water by a thick, whitish cream, produced from her skin and covering it like the coating of a channel swimmer. It is called vernix (Latin for varnish).

29–32 weeks

She weighs about 1400g and measures 40 cm. There is now relatively less water in the uterus, and it is easier to feel her arms, legs, head, and body through your wife's abdominal wall. She is still thin, but is starting to accumulate more fat. She cannot be harmed by pressure through the abdominal wall, such as when a doctor examines or when you have sex. (Your wife may however, find pressure on her abdomen too uncomfortable and you may decide together to use a different position for sex.)

Your baby is shedding the fine hair on her body, but the hair on her head is growing longer.

If born at 32 weeks, she nowadays has a more than 90 per cent chance of survival with good, special or intensive care. Most of her regulatory mechanisms are too immature to be left to their own devices. She may need the assistance of extra oxygen and perhaps artificial respiration on a ventilator. She is also liable to stop breathing and therefore lies on a mattress which sounds an alarm should that happen. Usually, just a small stimulating poke will start the breathing again. She may easily develop infection and need the assistance of antibiotics, as her own immature immune mechanisms may not give adequate protection from the normal germs in her environment.

33–36 weeks

Your baby grows another 5 cm and gains 900g. She is putting on the fat which, by reducing heat loss, helps to maintain her body temperature after delivery. Up until this time, your baby has been able to turn head over heals fairly readily, but now the amount of fluid relative to her size is diminishing, and she is more restricted. She usually finally settles with her head down, as there is more room for the larger buttocks and legs in the top end of the uterus. This means that for birth the head comes first, or presents, to the birth canal. This head-first orientation is called a head (or cephalic) or vertex presentation. If she becomes trapped, with the bottom end or breech coming first, we have a breech presentation. The likelihood of being able to correct to head first becomes less as time goes by, but some babies can manage it right up to the onset of labour. Sometimes your doctor may encourage the conversion by gently nudging her through your wife's abdominal wall.

The movements are surprisingly strong and if you are lying together you may be woken by a sudden firm kick—it doesn't necessarily mean you are unwelcome.

By 36 weeks, the greatest diameter of the presenting part (head or breech) may have moved down through the entrance to the bony pelvis (the brim of the pelvis). This we call engagement of the head (or breech). In some women this happens only while standing. Engagement may occur from 28 weeks onwards or maybe not until the labour is well advanced. It is usually later in second than in first pregnancies. Early engagement does not imply early labour and late engagement does not imply difficult labour. The main point of commenting on engagement is that with it one of the uncertainties of labour goes—whether the head is able to pass through this upper entrance to the bony pelvic canal. There is little point in you or your wife being concerned if engagement happens to be later than average.

37–40 weeks

Your baby is about to emerge—the process we call labour. This starts in 85 per cent of pregnancies between 38 and 42 weeks after the first day of the last menstrual period—this we call 'being on time'. This expectation, of course, only applies if the date of the last menstrual period is validated and if no subsequent information comes to light to cast doubt on the age of the fetus (*see* p. 33).

In the 266 days since conception, your baby has increased from the just-visible fertilized egg to 200 million cells, and her weight is six billion times that of the fertilized egg. Her rate of growth slows over the last weeks and there is often little change if her emergence is delayed until after 40 weeks. She is much more cramped in the uterus in these last few weeks, and therefore seems to move less, but she is strong enough to raise a lump on your wife's abdomen when she stretches. Bouts of hiccuping which may have been occurring from time to time during recent weeks tend to be more common. Most women describe this as a beating or rhythmical or pulsating movement at somewhere between 15 and 30 per minute—usually low down in the abdomen. It is not synchronous with the mother's pulse, and it is not the baby's heart which beats at 120–160 times per minutes. In the post-natal period when your wife watches the movements of hiccuping she will find them familiar—in the same way as she will recognize the limb movements which are characteristic of your baby.

Your baby is now nine months old and is quite tough and hardy. She has gained immunity to a wide range of infections by the transfer of antibodies from your wife's blood—the antibodies your wife has made through having had these infections or being immunized against them. They include poliomyelitis, mumps, measles, chicken-pox, influenza, and colds. Further antibodies are accumulated through breast-feeding. Colostrum, which is the yellowish fluid which comes from the breasts in late pregnancy and in the first few days before the true milk supply is established, is thought to be rich in these antibodies. These transferred antibodies are active in the first few weeks and months, and prevent or modify infections until your baby's own defence mechanisms are more active.

At birth, your baby's hair will be about 2 cm long and often forms quite a thick mop. Her nails protrude over the end of her fingers and toes, and she is well rounded. Average weight is 3 kg and length 50 cm, but the normal range is quite wide. Her eyes are usually a slate colour which gives little hint of their eventual colour.

It is easy to develop a rather romanticized image of what your baby will be like at birth. You may imagine her looking clean and tidy as

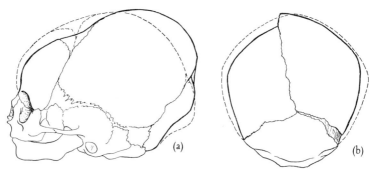

The moulding of the skull bone (a) seen from the side and (b) seen from behind with the bones over-riding

though freshly bathed. Birth isn't really like that—it is rather messy but none the worse for that. Your baby's head has a degree of 'plasticity' so that it can be moulded by the forces exerted during labour to a shape which most easily passes through the birth canal. The five flat bones which together form the vault of the skull are not rigidly fixed to each other as in the adult skull, but are connected by fibrous tissue to allow some movement. The bones can to some extent ride over each other, from side to side and from front to back. The shape of the head varies, depending on its attitude and position during labour, but with most babies the leading part is the top of the back of the head—called the vertex. The head tends to be elongated and more barrel-shaped from front to back. The elongated tendency is accentuated by swelling of the scalp where it protrudes through the cervix after rupture of the membranes. The part of the head protruding swells because it is not receiving counter-pressure from the cervix. The pressure from the edge of the cervix accentuates this tendency to swell. The swelling is called the 'caput succedaneum' or 'false head'—usually just 'the caput'. As the baby comes down to deliver, the first part we see and comment on is this swollen scalp. As the underlying bony head may not be down far enough to have separated the main muscles of the pelvic floor, progress may still be very slow, especially with first babies, even when we can see quite an area of scalp. Because of the moulding, the forehead often slopes back acutely and the chin seems to recede. The soft, cartilaginous end of the nose may be pushed over to one side so that it looks squashed.

Your baby will still have a covering of whitish cream (vernix), but less so if delivery is well past the expected date. Most babies have a blue tinge, as their oxygen supply tends to be rather reduced at the end of the second stage of labour. She may have passed bowel content into the liquor. As bowel content is a dark green, because of a dominance of bile, she may appear to have a green wash over her. Finally, there is often some bleeding from the fully stretched cervix at the end of the first stage, and there may also be bleeding from an episiotomy or tears at the vaginal opening. She is therefore often streaked with blood and blood-stained mucus (note that it is not your baby's blood, and she isn't bleeding).

If you are not expecting these various adornments and alterations of shape, you may get a bit of a shock and wonder which one of your wife's relatives looks like this! However, if you have thought about the newborn realistically and ahead of time, and realize that these appearances are normal, you will hardly notice them. You see your baby through it all as a wonderful and amazing new phenomenon—a complete little person! I have, by the way, painted the most unattractive picture, and would reassure you that most babies are really quite attractive at birth.

The length of pregnancy

Babies are born an average 266 days from the time of ovulation or 280 days (40 weeks) after the frist day of the last menstrual period, when the menstrual cycle has been recurring at monthly intervals. About 85 per cent of babies will deliver between two weeks before and two weeks after this date if the pregnancy is normal. This, therefore (between 38 and 42 weeks), is regarded as being 'on time'. If there have not been three regular cycles after a pregnancy or after stopping the pill, we cannot assume that the cycle pattern has become established and the date of the last menstrual period cannot be taken as a guide to the age of the fetus.

A varying length of menstrual cycle upsets the calculation. If the menstrual cycle length varies, say, between three and six weeks, then the expected time of delivery would be between 37 (38−1) and 44 (42+2) weeks from the start of the last menstrual period. Ovulation—the release of the egg—occurs about two weeks before the next menstrual period would have come if conception had not occurred. If the length of a menstrual cycle varies, then it is the part of the cycle *before* ovulation which varies. This means that ovulation and hence the age of the fetus cannot be predicted in these people by knowing the date of the last menstrual period. The menstrual period is really the end of the previous cycle, not the beginning of the next.

I apologize for mentioning one or two more confusing conventions in the hope that they may prevent misunderstandings.

The length of pregnancy can be expressed as the number of days from ovulation (which is much the same as from fertilization which takes place up to a day later). This is the convention often used in describing the age of the embryo during early development before the organs have formed.

Alternatively, the length of pregnancy can also be expressed in weeks from the first day of the last menstrual period. On this basis, pregnancy lasts 40 weeks. This convention ignores the fact that the fetus is really two weeks younger as calculated from the time of ovulation.

Finally, the length of pregnancy can be expressed in months—usually meaning calender months, in which case pregnancy lasts nine months plus seven days. Rarely, four-week (lunar) months are used, in which case pregnancy lasts 10 months.

Further information on the length of pregnancy is obtained by feeling the size of the uterus in the first few months with two fingers in the vagina and the other hand on the abdomen. Later the size can be estimated by feeling the uterus through the abdomen. An ultrasound machine can measure the crown–rump length of the embryo during the first 12 weeks of pregnancy, and after that the baby's head size can be measured. Measuring the fetus in this way becomes less reliable as a guide to the length of pregnancy after 28 weeks because of the varying sizes of the fetuses of the same age.

Very rarely, an X-ray may be used near the end of pregnancy when it is important to know the age of the fetus and there hasn't been an opportunity to collect this information earlier.

All these methods have margins of error and can sometimes be misleading, so that it takes considerable experience and circumspection to come up with a reasonable best estimate of the length of pregnancy (and therefore the age of the fetus) in difficult cases.

4 Guidelines for those contemplating pregnancy

When having a baby we tend to think nowadays very much in terms of the quality of the product. Doctors and nurses have paid a lot of attention to care and management during pregnancy and labour, and to care of the newborn baby. We are now in addition looking more and more at the time before conception and at the early weeks during which pregnancies are becoming established. We are turning our attention to the measures which may be taken *before* pregnancy to improve the development and growth of the baby. We are also planning ahead to optimize the smooth running of early child care and the early development of the family.

There is an important place for foward planning. However, this isn't everyone's approach to life and it can be overdone. Sometimes the decision to have a baby can be delayed too long, waiting for ideal conditions which may never be attained. The couple, initially keen to have a baby, can become too settled to contemplate the inevitable changes of lifestyle. A mutual decision to start a family is the theoretical ideal but sometimes it can be necessary for one or other to be assertive and to make the decision, or for both to throw caution to the wind and to take the plunge into parenthood. An unexpected or unplanned pregnancy may be no great tragedy for the couple who are adaptable and loving. Even though modern contraception can be quite reliable, accidents can happen and human error in using contraception is not unheard of!

You should encourage your wife to get into the habit of recording the dates of her menstrual periods so that she has the information in case an unexpected pregnancy occurs. A reliably documented menstrual history can be better than modern technology in deciding the length of pregnancy and the age of a fetus. Remember that the life and health of your baby may depend on this information. Many serious problems of late pregnancy, such as failure of the baby's growth due to suspected placental failure, can be resolved by delivering the baby at the stage when it is considered that its chances of dying in the uterus are greater than the dangers of its being born early. These latter are dependent on the age and maturity of the baby rather than his size. If a decision to deliver results in the baby's being born unexpectedly immaturely, then his chances of survival or of escaping brain damage in his early days may be significantly reduced because of his early arrival. I would like to see legislation requiring all women to keep an accurate menstrual calendar. Default would provoke severe penalties! Such a measure would make an obstetrician's life much more carefree!

Some medicinal drugs can have an adverse effect on early fetal development. As a general principle, the lower the intake of drugs of any sort in early pregnancy, the better, but there are some drugs, such as insulin, which can be essential for health. If your wife is taking any medicines and you have not already been advised about their safety if taken in early pregnancy, then consult your doctor.

As far as your contraceptive method is concerned, we usually advise that you switch from the pill to the sheath or a vaginal diaphragm for three or four months prior to intended conception. This is mostly to allow the menstrual cycle to re-establish for dating purposes, and slightly because of the possible and quite unsubstantiated anxiety that the pill could have some adverse effect on early development. Ideally, smoking and alcohol drinking should cease before pregnancy. It is possible, though still theoretical, that an adverse effect of tobacco or alcohol could be manifested through your sperm. There is certainly increasing evidence of the adverse effects of your wife's smoking during pregnancy. It is much easier for your wife if you stop smoking at the same time. You will get the lifelong benefits if you take this opportunity, and remember that the inhalation of your fumes by your wife or newborn baby may have a small though as yet unsubstantiated adverse effect during pregnancy or on your newborn baby.

There is clear evidence that a developing fetus may be seriously damaged both physically and mentally if his mother has a high alcohol intake during pregnancy—either continuously or as binge drinking. In addition, it is by no means clear that even a very modest intake is safe. Alcohol is a poisonous chemical not normally found in unspoiled natural food, and it is worth remembering that the delicate processes of early embryonic development are dependent on finely balanced chemical reactions. We now advise complete abstinence before and during pregnancy as the ideal. If a woman feels she must have a drink, then this is a good reason for stopping the habit, while if she does not have any strong desire, why take it at all, even if there is no certain risk? You can encourage your wife and be quick to defend her principle against 'friends' who so often press her to conform with group standards. You yourself can cut down or refrain to make it easier for her during this time, but I admit social pressures can be difficult to resist. However, non-alcoholic drinks are usually available.

If either of you has any doubts about health, possible inherited defects, or any unresolved worries, go and consult your doctor, preferably as a couple. Sometimes there are unresolved anxieties about a previous pregnancy problem. There may still be questions you wish to ask about previous pregnancies which have resulted in miscarriage, premature labour, or caesarean birth. It is much better to go back again and discuss in order to clear away lingering anxieties, doubts, or fears.

If your wife has diabetes, this is a good example of a medical disorder which needs attention. It should be managed with the most careful control before and after conception and, of course, during pregnancy. Until recently, the babies of diabetic mothers have had a two to three times greater chance of developing abnormally. There is now every likelihood that this problem can be significantly reduced by the most careful control of the diabetes before and following conception. The same applies to late pregnancy problems—they can be reduced by good control. You should make every effort to assist your wife in maintaining near-perfect control of her diabetes.

You may be worried about the likelihood of a recurrence of an inherited or developmental problem in an earlier child or in either family. You may be anxious because of your own or your wife's age. Read the section on abnormalities in the development of the fetus and then go along with your wife and talk to your doctor. He will refer you for specialist advice if it is needed.

Rubella (German measles) epidemics recur unexpectedly as a mild illness in the community, but if a non-immune woman develops the infection during early pregnancy, the fetus can all too frequently be disastrously damaged. There is a highly effective vaccine available, so any woman contemplating pregnancy who does not know herself to be immune should have a blood test to find out. If the blood test shows non-immunity, then she should have the vaccine and you should use reliable contraception for three months until the vaccine virus is cleared from her body. There have not been any reported cases of fetal damage in those women who have become pregnant soon after having the vaccine, but there remains a theoretical possibility as the attenuated virus of the vaccine is live, and multiplies and can spread to a fetus in the month or two after vaccination.

What about diet and physical fitness? Here again, we are not sure that a poor diet prior to or in early pregnancy increases the chances of fetal deformity, but this may be so. Rather than taking vitamin pills, it is probably better to have a good, well-balanced diet, which will contain these and other important food factors. To rely on additional doses of particular vitamins has not been proven to reduce fetal deformity and there is always the anxiety that unnaturally high doses of particular vitamins could in themselves have a harmful effect on early fetal development. If you and your wife get used to having a good diet before pregnancy, you are also well on the way to ensuring it during pregnancy and hopefully on a much longer term basis. If your wife is considerably overweight, there is a good case for weight reduction before embarking on the pregnancy. Significant weight reduction should not as a general rule be undertaken during pregnancy, as there is a possibility that dieting could upset the body's acid balance.

The diet we advise before and during pregnancy and during breast-feeding is very little different from the diet we would advise for those wishing to remain well and keep youthful weights. It should contain adequate protein, fresh fruit, and vegetables and should not favour fats and carbohydrates. Satisfaction in eating can be partly provided by taking the time to relax and eat slowly and to have your limited food intake attractively prepared and served. It takes a smaller amount of protein than of carbohydrate food to satisfy your appetite.

The carbohydrate foods to be taken sparingly are those rich in sugar and starch. They are the cheapest and will be offered everywhere you go. They are all forms of sweetened drinks and beverages, chocolates and other confectionery, sweet puddings, jams, honey, cakes, scones, biscuits, cereals, rice, bottled sauces and mayonnaises, prepared meats and thick gravies.

Fats are really concentrated fuel foods and are packed with calories. Their main value is to make food palatable. Unfortunately, most foods rich in animal protein are also rich in fat: meat, milk, cheese, some fish, eggs, butter, and cream. If you or your wife has a tendency to put on fat easily, watch out for these. It is better to seek lean meat and to grill rather than roast or fry. All fried foods are soaked in fat. It is better to use vegetable oil rather than animal fat in cooking. Butter (or preferably margarine high in polyunsaturates) should be scraped across rather than laid thickly on the morning toast.

Protein is one of the most important substances of all living matter, and food should contain a liberal amount of it. It is present in animal food products such as meat, milk, cheese, and fish and in many plant foods, like nuts, wheat germ, beans, and peas. A pint of milk (either as milk or in other beverages or in a sweet) an egg and a good helping of lean meat, fish, or legumes each day will supply plenty of protein.

Vitamins and minerals are contained in the protein foods already mentioned, but fresh, raw, or lightly cooked fruit and vegetables are principal sources. Try to develop the fruit and salad habit.

It is difficult to assess the value of exercise as far as pre-pregnancy and pregnancy are concerned. There is no proof that those who are physically fit are more fertile, but they are likely to have a greater feeling of well-being. The exercise habit is catching on and the very least that can be said for it is that, as an exerciser, you are likely to be in fashion! There is little doubt that fit women recover more quickly from normal deliveries and especially caesareans.

Planning the time for baby's arrival—easy in theory—is not always so easy in practice. Conception is most likely between days 10 and 18 from the start of a period for those women who have regular menstrual cycles (or lengths varying between 26 and 30 days). There is no need to be too concerned about the timing of sex, but if the intervals between making love are more than two to three days about this time, then it is easy to miss the time of ovulation and therefore the chance of conception. Sperm retain the capacity to fertilize for one to three days after inter-course, while the ovum can be fertilized within 24 hours of ovulation, and possibly for two or even three days thereafter. The sperm reach the outer end of the fallopian tubes within a few hours of intercourse.

Fertile couples take on average between three and five cycles to achieve a pregnancy, but may make it with the first ovulation. Of all couples, nearly 80 per cent have a pregnancy within 12 months and nearly 90 per cent within two years. Some of the remainder achieve a pregnancy without any sub-fertility treatment—sometimes after many years.

When should you seek medical assistance following failure to achieve a pregnancy (sub-fertility)? The simple answer is: at any time you suspect a problem or are worried. In general, it depends on your age and anxiety about pregnancy. For those over 30, after one year, and for any couple within two years. If at all possible, you should go along as a couple. There are often both male and female contributions to a fertility problem, and it is generally accepted that a male factor is present in up to 50 per cent of cases. Don't hesitate to ask question, and make sure you understand any explanations or instructions you are given.

Think and talk about how your lifestyle may be affected by your wife's pregnancy and by your baby and child. Many aspects of your life will be enriched and will take on new meaning, but there will be restrictions. If you have tended to have separate interests and leisure time pursuits, you may find that your wife quite reasonably wants more of your attention during pregnancy. When the baby arrives, you are likely to be very much on parade, helping with the baby and household chores—there will be little time for cricket, and your friends may think of you as lost for ever!

5 Diagnosis of pregnancy and arrangements for delivery

Diagnosis

Your wife will first notice that her expected period is late and, if she gets premenstrual symptoms such as breast tightness and soreness or a heavy aching feeling low in the abdomen, these can persist without the expected period materializing. After about a week, she may notice nausea or even vomiting at any time of the day or evening, but perhaps most likely in the morning. This symptom can appear even before the period is missed. Some women just feel different, while others have no alerting symptoms. They feel normal.

If your partnership has been sub-fertile and your wife has been keeping a record of her early-morning temperature to check on the time of ovulation, the elevation of temperature which follows ovulation persists. It doesn't drop down again as it does when menstruation is about to start. Make sure your wife keeps this temperature chart and shows it to the doctor at her first visit, as it usually allows accurate determination of the time of ovulation and therefore of the age of the pregnancy.

A pregnancy test can be carried out by your local clinic or chemist, or your GP, or you can use a home pregnancy-testing kit. The first specimen of urine passed in the morning is used. Unfortunately, the test, although quite accurate when the period is nearly two weeks late, cannot be completely relied on, even when the following points are considered. The urine must be caught into a bottle, which is *laboratory* clean. No matter how carefully you think you can wash a receptacle, the remaining faint traces of impurity or detergent can upset the test. Medicines which your wife may be taking can also interfere, and the experience of the person carrying out and interpreting the test is another factor. If the test is positive, there is a small chance that your wife is not pregnant and, if negative, she can still be pregnant. If this is so, you may well wonder: why bother? I also wonder, but know that many people like to use these tests. Provided that you take the result of a home pregnancy-testing kit and perhaps that of your chemist with a few grains of salt then I guess it's harmless. There are very accurate new tests which use either urine or blood. They will replace current tests. If your wife has irregular or infrequent cycles, then it is hard to know when to suspect pregnancy if she also is one of those who doesn't have any of the symptoms of early pregnancy.

The visit to your doctor

You may like to go with your wife, as this is the time to discuss with your doctor arrangements for antenatal care and delivery. Go as soon as you want to, but certainly by the time of the second missed period. Currently new tests (chorion biopsy) for the very early diagnosis of fetal defects are being developed. In future years the advice will therefore probably be to report as soon as pregnancy is suspected. Hence new tests.

Home or hospital or general practitioner unit?

Home delivery is still chosen by a few women in Britain but it is interesting to note that, even in Holland, which has maintained a strong tradition of home delivery, more and more are opting for the greater safety of hospital. The debate is by no means simple. There are various statistics available which seem to prove the greater safety of hospital delivery for mother and baby. Others point to the safety of home delivery when there has been careful selection of cases, good obstetric and paediatric flying squad back-up, and the exclusion of cases who accidentally deliver at home. Undoubtedly, the ones most at risk in most ways are those who intend home delivery but need transfer to hospital during the course of labour.

As so often happens, we have in the end to get down to individual cases. Sometimes the women who try to arrange or insist on having home delivery are the very ones who should not, because of easily recognizable factors which increase the risks. First deliveries or fourth and later deliveries, for instance, are regarded as of unacceptably high risk. Even when low-risk women are carefully selected, the unexpected can materialize, even though this is unusual. For the individual who is confident in the home setting and has a relatively straightforward labour and delivery in familiar surroundings, home delivery can turn out to be ideal. For the occasional case which goes horribly wrong, nothing can be more depressing—especially if the problem is one which could have been resolved efficiently with the facilities and help available in the hospital.

Sometimes a couple opt for home delivery much more because of the negative factors, real or presumed, associated with hospital management—strange and frightening surroundings, apparatus, un-friendly people, absence of personal continuing care. Undoubtedly, these disadvantages are sometimes real, but there has been a vast overall improvement in humanizing the arrangements and management in many hospitals. There is more attention paid to the requirements and needs of individuals. Comprehensive antenatal education programmes can build up confidence, trust, and familiarity. The security of the safety element provided in hospital can greatly relieve anxieties—in other words, the hospital can score better on the very aspect which is sometimes said to be lacking, namely the lowering of anxiety.

Trust in the personal care of home confinement can be shattered if labour occurs at a time when the midwife or doctor is off-duty and a stranger appears. When nothing seems to be happening in labour, the midwife can slip off to attend to some other problems, only to find that delivery of her multigravid patient has occurred unexpectedly in her brief absence. Because of the small numbers of home deliveries in an area, the flying squad arrangements may not be kept well oiled for the few occasions on which they are needed.

A very good compromise is the general practitioner delivery unit, where the same low-risk cases as can be considered for home delivery are managed by the general practitioner and perhaps the group practice midwife. These units are ideally situated in hospital adjacent to a consultant unit, usually in the same building. Safety and personal care are then likely to be combined. In other units, the GP has his or her patient managed within the consultant labour ward, often in associa-tion with the district midwife. Other combinations exist. The advanta-ges are that your wife sees her own GP for most of her antenatal appointments and builds a relationship with him/her as well as with the midwife who will probably be the person actually delivering her baby.

You should think about these alternatives and talk them over with your family doctor. He is likely to know the detailed workings of local arrangements.

You may have heard the term 'birthing centre'. As usual, this seems to mean different things to different people and is sometimes more a new label than an entirely new concept. The idea is to emphasize the home-type environment and to demedicalize the atmosphere by having any equipment or back-up services out of sight but readily to hand if needed. In some places the couple may have family or friends at the birth. The woman and her baby (and husband) stay in the delivery area until they go home, usually within 6–12 hours of delivery.

If you opt for a hospital consultant unit delivery after being advised by your doctor, there may be more than one within reasonable distance of your home. Try to find out from your doctor, friends, and acquaintances the points for and against each. Try to find out about a particular hospital from people who have actually had their baby in it relatively recently. Seek several opinions, for individual experiences may not be representative. Don't be too influenced by rumours as these are sometimes perpetrated by those with no first-hand knowledge, only lively imaginations. Beware of needing to travel long distances in labour if you are having a *second or subsequent baby*. The unpredictably rapid labour and then delivery on the way can prove too much for the nerves! Sometimes if these distances are unavoidable, an elective induction (*see* p. 173) is justified, or your wife may for the last week or two move in with a friend or relative closer to the hospital.

Breast- or bottle-feeding

You may be asked early in pregnancy whether your wife intends to breast- or bottle-feed your baby. Of course there is no need to make a final decision at this stage, but you may like to read the relevant section in Chapter 19 (pp. 150–4).

6 How will you both feel during pregnancy?

Pregnancy is a time of emotional change and reorientation, which continues after the birth as you become a family of three with a somewhat different outlook and different priorities. This means that you, as well as your wife, are going to adjust and perhaps make some sacrifices. Pregnancy and early parenthood can be a time of crisis or disequilibrium. Your relationship is to some extent in transition between the romantic love of early marriage and marital love which encompasses more responsibility and a third person. You and your wife may find difficulty with this change because you expect your love relationship to remain much the same despite the changes of pregnancy and parenthood, and the passage of time. It is perhaps more helpful to keep your mind open to the maturing consequences of your new life situation. You may be surprised to find yourself becoming more responsible in material things and being a little more conservative and future-orientated. Often a woman has remarked that, whereas her husband had been relatively carefree and incautious about money, she notices during pregnancy a much more responsible concern to get organized for the future and to pay off debts.

Some men worry much more than usual, have sleepless nights, and become depressed more than circumstances should dictate. Try to talk with each other about the way you both feel, but if either of you finds that you are not really coping, don't hesitate to go and see your doctor or at least talk with a trusted friend. Although it is relatively rare, real depression can become an unshakable burden.

You may also notice that your wife becomes more dependent—she will want you at home and with her more of the time. She will try to get you to be more careful, for example when you drive. She will worry much more when you are late home, and will easily get into quite an agitated state. Try to be even more considerate than usual—send a message if you are delayed. Be prepared to forgo now and then your regular get-together for a few drinks at your favourite local.

On the other hand, it is important that you don't become socially withdrawn as a couple—especially when your wife stops work. Encourage her to continue to call in at her work place. Make sure you both keep seeing your friends. If you have any friends with a baby or young children, it is a good idea to build up baby-sitting credits which you can call on later—you also get practice for your future role.

You will probably enjoy seeing your wife change physically as your baby grows. Many men find these changes fascinating and very attractive. However, don't forget that during her pregnancy your wife may become unsure of herself and of your affection. She may worry that you will find her new shape unattractive or that your eye will wander to other women. Talk about the physical and emotional changes of pregnancy, and try to understand them, and, above all, be reassuring. Rather than take it for granted, tell your wife how attractive you find her and how much you love her. Your relationship will thrive on extra cuddling, caressing, fondling, and sexual pleasuring. Your wife may need this proof that you still love her and find her attractive. You may be puzzled when at times she is happy for this intimacy and yet doesn't want to go on to full intercourse.

Sex during pregnancy

Sex during pregnancy is, as at other times, much more than just the act of sexual intercourse. With many couples, the frequency of sexual intercourse is lessened in early pregnancy and again towards the end of pregnancy. It is not surprising that a woman who is feeling tired, nauseated, and 'out of sorts' is uninterested in sex in early pregnancy, but her inclination usually returns as this phase passes. Self- and mutual stimulation sometimes tends to increase as the frequency of intercourse diminishes, whereas for others, pregnancy can be a time of much greater sexual indulgence. Some women go off sex altogether and some men can have all sorts of worries and anxieties about the consequences of sex during pregnancy. Some couples can, especially in late pregnancy, be concerned about the invasion of their privacy by a third person. The idea that the baby is present nearby in the uterus during intercourse can take a bit of getting used to!

As far as the effects of sex in pregnancy are concerned, all that is known is that, with one or two possible exceptions which I shall mention, there should be no medical discouragement of sex. You and your wife should try to discuss your feelings about sex as openly as you can. Talk over your anxieties and wishes, remembering that you as a couple are individual and unique and won't necessarily be the same as most other couples. If you have continuing worries or anxieties you may be able to find the information you want from books, but don't hesitate to go along to an antenatal clinic and ask about any remaining queries, or encourage your wife to ask on your behalf. Continuing anxieties about sex should not be allowed to upset your relationship.

There is no evidence that sexual activity of any sort is responsible for miscarriage. The developing embryo (or fetus) and the placenta are safely enclosed within the uterus which is suspended from the side-walls of the pelvis and freely mobile. Neither the thrusting of the penis nor pressure on the abdomen will affect them.

Men have expressed concern about their penis denting the baby's face, or about semen washing over the baby's face. The baby is tucked away inside the uterus, floating in water, and is separated from the semen deposited in the vagina by the cervix, the mucus plug, and then the membranes, i.e. right out of contact!

At the time of sexual climax (orgasm), however produced, the uterus does contract more strongly, but there is no evidence that this predisposes to miscarriage. Miscarriage is common and it is not surprising that when couples look back for a possible cause, intercourse a few hours or days or weeks beforehand tends to get the blame! There is normally no reason for taking the pregnancy into account when contemplating intercourse. Despite this reassurance, some couples who have already experienced a miscarriage remain concerned that intercourse could bring on another miscarriage. This concern is more on the emotional than the intellectual level and they decide to refrain from intercourse during the first three to four months. The important thing is to talk to each other about such fears. Remember that your wife may be much more concerned than you suspect and she may for your sake be trying to suppress her fears. Much better that you ask how she feels and talk about it.

There is a very widely held myth that miscarriage is more likely at the time of the missed periods and that great care should be taken at these times. This is quite untrue, as the statistics of the frequency of miscarriage show no increase at these times. In addition, I cannot think of any reason why this should be so. The whole of the menstrual cycle is in abeyance during pregnancy so that the body mechanisms wouldn't have a clue as to when the missed periods would have come!

Sexual arousal, especially the formation of lubricating fluid, may come less readily during pregnancy. Without normal swelling and lubrication, penetration can more easily abraid the softer tissues around the opening of the vagina. The opening of the urethra close to the vaginal entrance is one of these tissues which is more easily bruised. This can cause greater frequency of urination or burning pain or stinging when passing urine. These symptoms are similar to those of a urinary infection and can mislead your wife into thinking that she has developed such an infection.

It is often a good idea to take more time with sexual pleasuring and arousal, and occasionally to use a little lubricant, such as KY jelly, if necessary, as dryness can, in some women, be partly an effect of the hormones of pregnancy.

You may worry that your penis is going to damage the baby's head, or that it will break the waters. These are especially likely to be anxieties later in pregnancy when the baby's head may be lower in the pelvis. None of these could happen, as the penis stays in the vagina and is directed back in the pelvis, below the uterus. The opening into the uterus is very narrow and would not be disturbed by your penis. Your wife, however, may find deep penetration too uncomfortable at this time. The watchwords for intercourse in pregnancy are generally patience, gentleness, and tenderness. Experiment with different positions, especially if your wife finds pressure on her abdomen uncomfortable in late pregnancy. (This, by the way, wouldn't harm the baby as pressure is distributed by the liquor.) Use plenty of pillows if they are helpful. Try lying sideways facing each other or with you behind her. Your wife may sit astride you so that she can control the depth of penetration. If your wife is on all fours the uterus tends to move out of the way. You may find approaching the vagina from behind in this position a satisfactory variation. Pregnancy can be a time of experiment to find the ways that suit you both at the time.

Never blow air into the vagina during pregnancy as this has caused sudden death of the woman when air has entered her blood stream via the blood vessels supplying the placenta.

Trying to bring on labour by having intercourse when the baby is due or overdue is probably a bit hopeful, but it is at worst a harmless experiment. However, if there is any question that the membranes may have broken and that the liquor is leaking, then penetration should not occur, for there is then a definite risk of infection being introduced. Always report to your medical or midwifery attendant if you think the waters may have broken.

If your wife has had an episode of threatened premature labour in this pregnancy, or has delivered early in the last pregnancy, or has a history of no normal pregnancy after a late miscarriage, I would advise that between 20 and 35 weeks of pregnancy you refrain from intercourse or sexual activity leading to your wife climaxing. As I mentioned earlier, we know that orgasm is associated with strong contractions of the uterus, and it is just possible that these contractions could trigger premature labour in a woman predisposed to it. You should refrain from sexual intercourse between 25 and 35 weeks of pregnancy if your wife has bleeding during this time, even though placenta praevia (see below) has been excluded. In a woman who has had bleeding from a normally situated placenta, premature delivery is more likely. It is possible that sexual climax could accentuate this tendency.

Placenta praevia is the final concern as far as avoiding intercourse is concerned. Placenta praevia means that the placenta is attached on the lower wall of the uterus near its opening. As far as intercourse is concerned, placenta praevia is most important when the edge of the placenta comes right to the edge of, or actually covers, the opening into the uterus. if this is the situation, then there should be no intercourse from about 25 weeks until the baby is delivered. Your wife will have been told if she has placenta praevia. It is possible that intercourse could cause a partial separation of the placenta if it is in this site. This could lead to serious bleeding. Nowadays, many women have an ultrasound scan at abour 15 weeks, and at that time an attempt is made to identify the site of the placental attachment. Sometimes it can be shown to be right across the opening. In this case, the above ban would operate. Much more often, however, all that can be said is that the placenta is 'low-lying'. The ultrasound scan is then repeated at about 32 weeks when, with the growth and expansion of the lower part of the uterus (lower segment), the placenta will usually appear to have moved away from the opening. In other words, the earlier suspicion of placenta praevia has proved to be apparent rather than real. A few of these earlier low-lying placentas will be shown at the 32-week ultrasound to be the less extensive varieties of placenta praevia. This is only of importance from then onwards. An early ultrasound diagnosis of low-lying placenta should be ignored as far as sex is concerned until the 32 weeks scan confirms placenta praevia.

Finally I think you will appreciate, from what I have said, that for practically all couples sexual intercourse in pregnancy has the green light.

Travel in pregnancy

Pregnancy is not the best time for travel because of the unexpected problems which may arise when you are away from your usual medical and midwifery attendants. The principal concern is the dislocation rather than direct effects of travel on the pregnancy. There is one very important point to consider—the increased tendency in pregnancy to form clots in the deep veins of the legs (deep vein thrombosis). Although these clots are rare, they are potentially very dangerous as they can travel to the heart and perhaps on to the lungs and block the circulation, with serious and even fatal results (pulmonary embolism). Your wife has to be wary of situations which slow the circulation through the legs—especially the return of blood in the veins.

On car, bus, train, or air travel your wife should wiggle her toes and tense up her leg muscles from time to time to speed the circulation. She should get out of her seat at least once each hour and walk around a little. On a car journey you must stop at least once each hour and insist that she gets out. It is so easy for her to get comfortable and not want to move about in this way, but it is unwise to maintain these cramped positions for long periods. If you have to travel a long distance, it is better to break the journey if this can be arranged.

Dehydration can also predispose to this problem. Your wife should drink plenty of fluid (and not alcohol) during an air trip or if visiting an area with a hot climate. Water is probably the best fluid provided it is potable. (Use boiled or bottle water if there is doubt.) Alcohol increases dehydration and of course it should also be avoided anyway because of its possible harmful effects on the fetus.

Nausea and vomiting of early pregnancy are often worse during travel so try to avoid travelling if your wife is having trouble with these symptoms. Travel-sickness pills can help, *but first check with your doctor that they are suitable to be taken during pregnancy.* Your wife should not take these tablets if she is doing the driving. They reduce concentration and reaction times, especially in late pregnancy. When she drives your wife should continue to wear her seat belt as this is in general much safer.

Check on the available medical services in any area to which you are contemplating travelling. If the services aren't first class, remember that any woman can deliver a premature baby whose whole future can be jeopardized by poor care in the newborn period. I would recommend that you return by 26–28 weeks rather than be on holiday from that time.

Make sure that your medical insurance covers problems relating to pregnancy. Without it you could be paupers for life if an expensive problem arose in a country such as the USA. Most medical insurances issued for travel do not cover pregnancy problems. If you are to be in a Common Market country take document E111 with you (available from your local Health and Social Security Office). Those vaccinations or innoculations which may cause a high fever or which involve a live virus are best avoided during pregnancy and a medical certificate will suffice to gain exemption.

If you are travelling to a malarious area you should both (but especially your wife) be taking prophylaxis tablets. (The recommended tablets vary with the area to be visited.)

Most airlines won't take your wife after 32 weeks without a medical certificate which absolves them of responsibility. I can see their point of view—as it can be difficult to provide emergency medical treatment in the air!

Your wife should not travel in a non-pressurized plane and if there is a real problem about the growth of the baby, the usual pressurization at 4000–5000 feet is probably an adverse factor. It would not, however, affect a normal pregnancy. If you want a short holiday then 18–23 weeks is the safest and most comfortable time.

Early pregnancy

We will go back a little to the start of pregnancy. Reactions to the news that your wife is pregnant can be puzzling. Even when you have both been looking forward to starting a family, the first news can bring a panic reaction. Your wife can suddenly see all the worrying consequences of pregnancy, and may doubt the wisdom of the decision for a pregnancy. Instead of elation, she may have a whole series of doubts, and will be anxious to see how you react. The rather startling news of a pregnancy may come as a bit of a shock to you, especially if it hasn't been planned. You can so easily give the impression of unconcern or even uninterest, and finish up in the doghouse, in disgrace! Do try to be positive and to appreciate the exciting and bright side, even if you don't feel fully committed. The news should at least call for a celebratory glass of real orange juice!

Nausea and vomiting during early pregnancy can be real dampeners and, along with these or instead of them, you may have to put up with your wife's peculiar cravings for say, strawberries and cream, or fish and chips, at 3.00 a.m! So-called morning sickness *is* more common in the mornings, but can be a problem at any time of day. Your wife may feel sick from normal cooking smells and can develop an aversion to the kitchen. This is where you have the opportunity to demonstrate your cooking prowess, or at least your consideration by being content with prepared foods. Encourage your wife to eat little and often and not to rush about too much. Keeping up a steady intake of nourishment seems to help. Encourage a snack before retiring and have a glass of milk, and perhaps a biscuit, at hand when your wife wakes during the night. Try to get her to stay in bed in the mornings until you have brought a cup of tea and dry toast or whatever she fancies. Having tried all this, some women continue to be sick and feel pretty miserable as a result. Your wife may need pills from her doctor or even admission to hospital and, rarely, an intravenous drip. The baby is not harmed by her sickness. Morning sickness usually passes after 12–16 weeks, but occasional women continue to be troubled for most of the pregnancy.

Mid-pregnancy onwards

In theory, women are supposed to be happy and contented and to look and feel really well in mid-pregnancy. Certainly most do tend in that direction but many, for no accountable reason, feel unwell or just below par for most of the pregnancy. Pregnancy affects different people differently. Having a distressing or uncomfortable pregnancy in no way implies a difficult labour and delivery, unless there is a definite medical problem.

There is usually much more emotional lability in pregnancy. Your wife's moods tend to be less predictable and to swing from one extreme to another. Sometimes she will fairly suddenly change from being quite even-tempered to feeling sad, burst into tears, and not really be able to tell you what's wrong. She may be much more sensitive to criticism and read far more into your chance remarks than you had anticipated. At other times she will be much more light-hearted and cheerful than circumstances seem to justify. You will learn to rub along with these moods, especially if you are both able to discuss them *at appropriate times*. Above all you should both try to retain your sense of humour. Disturbed emotions during pregnancy in no way imply any difficulty for your relationship as a couple, or for your abilities to relate well to your baby and to make good parents. Some women in early pregnancy and again in late pregnancy become excessively tired. Your wife may become absent-minded, vague, and forgetful.

Some women who remain well can manage to stay at work quite happily until late in pregnancy, while others find the effort too much. The nesting instinct often seems to operate strongly and you may find it distinctly hazardous to suggest moving house towards the end of pregnancy! It is better to effect any upheavals early in the pregnancy. Your wife would rather you concentrate in the later part of pregnancy on preparing the nursery and on making your home more childproof. You will be contributing to nest-building. One of the frightening realizations which dawns on women in late pregnancy is the inevitability of the process. They can feel trapped by being obliged to go through with labour and having the baby—they are finally committed to having no control over this aspect of their lives.

You too may find yourself less emotionally stable during pregnancy than your self-image dictates. There is a fascinating phenomenon called the couvade (from the French verb *couver*—to brood or hatch). Some ten per cent of husbands develop, often from early on, some of the many symptoms of pregnancy. These fade around the time of the safe arrival of your offspring. The 'popular' ones centre on the various forms of indigestion, heartburn, loss of appetite, constipation, and backache.

With an increased appetite you may well find yourself putting on weight, with a prominent belly made more noticeable by the way you adopt the sway back-type posture and rather waddling gait sometimes assumed by pregnant women. I have known quite a few men who developed the habit of passing urine frequently during the night (in sympathy?). Some observers have suggested that these various symptoms occur more often in men with a more worrying, anxious type of personality. No doubt they at least reflect concern for the wife's welfare.

Some women develop quite a strong feeling of attachment for their babies from mid-pregnancy onwards, while others don't relate much until perhaps they notice movements lessening, or the baby having a quiet period. Naturally enough, a woman can soon become worried. Babies do have quiet periods, but if after 30 weeks there are less than 8–10 noticeable movements over a ten-hour period, then it is as well to report to your doctor. Usually all is well, but sometimes an electrical recording of the pattern of the baby's heart will be taken. Very rarely, this may give an indication that it is time to deliver the baby even before he is due.

One of the objects of the antenatal clinic is to allow the early detection of any problem which may be arising. It is not surprising, however, that your wife will be very upset if, when visiting the doctor, it is suggested that she be admitted to hospital for a period of observation —usually because her blood pressure has been found raised. This by no means spells disaster, but it can be the start of a blood-pressure problem. Most times, frequent measurements over 24–48 hours show it to be a false alarm, and your wife returns home. She will usually need plenty of reassurance during such an episode, and you may also come into the same category. You can't very well reassure your wife unless you understand the situation yourself, so do ask questions if you need to. An alternative to admission to hospital for the assessment of lesser problems is for your wife to be asked to remain for several hours for day assessment.

Planning for the future

Pregnancy is the time to start discussing and thinking about child care and child-rearing. Remember that you have each come from a different family in which attitudes and practices in child-rearing are likely to have been different. The natural tendency is to revert, in the care of your child, to methods and attitudes with which you were nurtured. Try to compare and contrast your individual backgrounds so that you can talk through your differences to find common ground. You need agreement on general principles to prevent the major discord which can spring from unresolved differences in the way you wish your child cared for. Talk about baby feeding and whether it is to be breast or bottle. For successful breast-feeding your approval, encouragement, and support are vital.

Your wife may wish to keep contact with her career and, especially if she is going to return to it on a full-time basis, you will need to do plenty of planning. Your wife will have to delegate most of the traditional household tasks to paid helpers, unless you are particularly fortunate in having unusual family arrangements. You will need to do half the child-rearing, or your wife will be attempting to achieve superwoman status and in the end fall short both in her career and as a parent. It is as well to retain flexibility in your planning, as your wife's aspirations may undergo considerable modification when she experiences caring for a baby and the emotional pulls it can involve. It is easy to grossly underestimate the difficulties you are likely to encounter in her maintaining her job.

Late pregnancy and delivery

The last few weeks of pregnancy can be a bit agonizing while you wait for labour to start. Remember that labour is on time if it starts between 38 and 42 weeks from the first day of the last period if your wife has had 28-day cycles. If cycles have been of 35 days then up to 43 weeks is still on time. Even if you consider yourselves part of the anti-induction lobby, it is so easy for you and your wife to start agitating for an induction of labour when there is no indication of labour beginning. It is of course one of the jobs of your professional adviser to resist this agitation unless the womb is so ready for labour that induction won't fail.

Make sure you have made preparations for your journey to hospital. Always keep petrol in the car and in winter make sure it is likely to start at 3.00 a.m. on a cold night. Keep the hospital telephone number handy and, if this is not your first baby, make sure the details for management of the other children have been worked out. Have a notice written to leave on the windscreen of your car, even if you arrive at night. 'Emergency. Wife in hospital in labour. Will be back soon—I hope!' Most traffic wardens will be sympathetic—anyway it's worth a try.

You should both have thought about the time after delivery. You may have arranged to bring in a bottle to celebrate, and it's a good idea for you to have alerted friends or relatives that you will be bringing the glad tidings in person perhaps at an awkward time. They will understand and want to assist your celebration. It can be a bit of a let-down to go home to an empty house by yourself when you are excited and want to share your experience, happiness, and news. The same applies to other nights when you come home after visiting your wife in hospital. Friends realize that this can be a lonely, sometimes worrying and frustrating time for you, so don't hesitate to accept their support. (And let's hope they give you some!)

7 Antenatal care: tests during pregnancy

The intentions of antenatal care are to observe the progress of pregnancy, expecting it to be normal but to detect as early as possible any deviations from normal. They are detected by regular observations, history-taking, examinations, and appropriate tests. Sometimes these deviations can be corrected. An example is the treating of a urinary infection, which is more likely to appear in pregnancy. Observation can help to decide the time when intervention is required, for example the early delivery of a baby which is failing to grow satisfactorily and therefore likely to be in danger of dying in the uterus. Antenatal care is often shared between the hospital clinic and your GP, or the Local Authority clinic, or a clinic run by a district or hospital midwife.

A number of special tests are used routinely, or more or less routinely, while others apply only when special situations arise. A few measures,

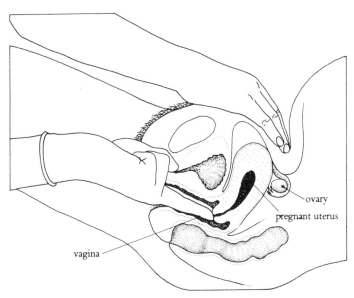

Pelvic examination at the first antenatal visit to check the ovaries and to check the size of the uterus

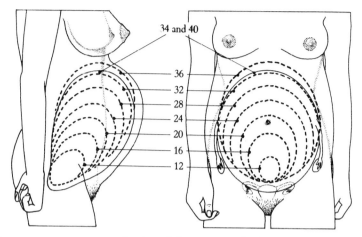

As pregnancy progresses, the size of the uterus is checked against this standard

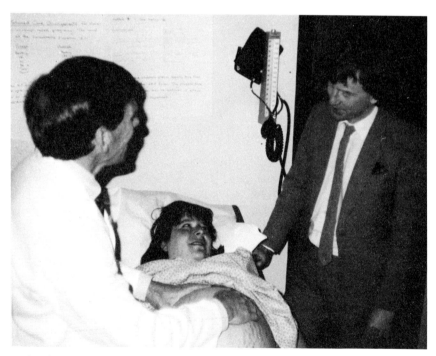

In the clinic you can take part in the discussions: in this case about the reasons for a probable caesarean section. (The stretch marks fade in 9–12 months.)

such as the prevention of anaemia by taking iron tablets, may be employed more or less routinely, while treatment for 'heartburn', varicose veins, or 'piles' is called for occasionally.

Antenatal education is not only the concern of arranged classes, talks, and lectures, but is a continuous process. This is partly achieved by giving information in response to questions, and partly by describing and explaining things as they arise, or discussing the pros and cons of a particular management.

Rather than attempt a comprehensive account of antenatal care, I shall pick out a few topics which are likely to interest you. These are mostly the more recently developed tests used to detect developmental abnormalities.

Developmental abnormalities

Before we dwell on this rather depressing subject, it is worth remembering that almost all babies will be born normal, will thrive, and will grow up to be normal adults. However, all potential parents, not unreasonably, worry that they will be the unfortunate ones. Sometimes there is concern about a problem which runs in the family—in such a case, or when you have already been the parents of an abnormal baby, you will be referred to an expert in this field (a genetricist or genetic counsellor), if your family or clinic doctor or obstetrician cannot give you full information on the subject.

The processes of development of a baby from the minute full-stop-sized fertilized ovum are so complicated that one cannot begin to understand how the finer details of organization and development are carried through. One could well wonder why any baby ever develops normally, rather than why this process occasionally goes wrong. Unfortunately, a slight hitch in an important early step can lead to a major defect in the baby. We know some of the influences which can contribute to some of the defects. Increasing maternal age is for instance a factor in the development of a Down's syndrome (mongol) baby. Rubella (German measles) may be responsible for eye, brain, or heart defects. Many influences are unknown and are perhaps often reflections of the inherent deficiencies in any natural biological process. Developmental defects often end in miscarriage or occasionally stillbirth—a few result in a seriously deformed baby who is more likely to be born pre-term or small for dates. Such serious deformities tend to lead to death in the first few days, weeks, or months, leaving only a very small number to survive with deformities which cannot be cured or satisfactorily treated by modern methods.

All parents worry that they may have an abnormal baby surviving without the potential to lead a reasonable life- -usually because of mental defect. There is no way you can be entirely freed of this worry until you have seen your baby and she has been declared normal. However, it can be helpful to keep telling yourselves that almost all babies are normal and healthy at birth.

In addition to babies who develop abnormally, there is another group whose defects are of about the same importance. These defects arise in babies who have developed normally but who are damaged as a result of problems arising just prior to birth, during labour and delivery or in the early new born period. Being born well before full term is a major factor here, but I will take up the problems of damage in Chapter 16.

Tests in pregnancy

There are a few tests which can be carried out by way of general screening for commoner defects. Additionally these tests can be applied in specific ways for a whole host of rare problems which are relevant for a few women. If your wife wishes to take advantage of these tests, if they are applicable in her case, then she should do so. You then have done all you can to ensure that your baby is normal. It is then better to look on the positive side and try to concentrate on the much more than 99 per cent chance that your baby will be developing normally.

The tests I describe are not really applicable if you and your wife would not contemplate having a pregnancy terminated when a suf-ficient abnormality is demonstrated in your developing fetus. Most people who feel this way would not have the tests.

The alpha-fetoprotein (spina bifida) test (AFP)

This test aims to detect errors in the way the brain and spinal cord have developed, but sometimes detects other rarer serious and less serious defects. Alpha-fetoprotein is a protein in quite high concentration in the fetus—it is normally contained within the fetus, but when the fetus is defective can 'leak out' at a greater rate and find its way into the liquor (waters) and into the mother's blood. It is normally present in small concentrations in the mother's blood, but occurs in a much higher concentration when a fetal defect is present. The best time for a sample of your wife's blood to be analysed is between the sixteenth and eighteenth weeks of pregnancy. The test can give a falsely high result if there has been bleeding during pregnancy, or if the pregnancy is in fact further advanced than has been calculated from the dates and physical examination. Sometimes no good reason is found for a confirmed high reading.

If your wife is called for a repeat blood test, don't be too worried as this is just the next screening step for a raised alpha-fetoprotein level which may warrant further investigation. Despite this advice, you will, of course, worry and we recognize that anxiety which later proves to be unfounded is a price to be paid if the diagnosis is to be made where there is a real problem.

The next step is for a full ultrasound screening to be carried out. This excludes twins and checks on the age of the fetus. The rest of the fetus is carefully checked for evidence of a defect of development.

If none is found, then it is quite usual to take, by amniocentesis, a sample of liquor for analysis for its alpha-fetoprotein content. At the end of all this, your wife's fetus or fetuses may be given the all clear; a major defect for which termination of pregnancy is offered may be diagnosed; a less serious defect may be present; or there may continue to be some doubt for which follow-up ultrasound screening is advised. You will see that none of this is simple or straightforward, and the judgement of experts is needed to advise and guide you.

At the time of writing there is no general agreement that the results of the alpha-fetoprotein test justify its use as a routine measure. It is not available everywhere.

Ultrasound screening

This may be initiated for reasons other than an abnormal alpha-fetoprotein level. It may be the basic screening method used in your particular hospital. It may be undertaken because a previous baby has suffered a particular developmental defect; an unusually large uterus containing an excessive amount of liquor may raise the possibility of an abnormal baby; or even growth failure of the fetus may require further examination. These last two are problems of later pregnancy.

The technique has, during the last 20 years, become extensively used in obstetrics, and allows us to visualize and measure the contents of the uterus, and to see how they are working. It is used to guide fine needles and telescopes to obtain specimens of liquor, or of placenta or of baby's blood for special diagnosis. It can also help in the use of instruments within the uterus to treat occasional special conditions such as blockage of the outlet of the bladder of the fetus.

Pulses of very high frequency, but extremely low-energy sound waves are passed into the uterus from the surface of the mother's abdomen, which is covered with oil or jelly. The sound waves are reflected back from interfaces between tissues of different densities. The returning patterns of sound waves are integrated by a highly sophisticated piece of electronic apparatus and displayed on a screen. The images obtained show the uterus, the placenta and its site of attachment, and the various parts of the fetus including most of the internal organs. The amount of detail seen depends on the type and quality of the machine, and the skill of the operator. Obviously all these investigations are time-consuming and expensive in terms of money and manpower. Local facilities may allow only a limited range of ultrasound tests and investigations. Sometimes only one or two centres are able to undertake the most complicated investigations.

Many hospitals now do at least one ultrasound examination at about 16 weeks to check that the size of the fetus (and therefore presumed age) corresponds with that expected. Twins are excluded and usually the site of the placenta noted.

If your wife is to have an ultrasound examination, try to go along with her as you will be fascinated by what you see. If a problem is suspected, you will be on hand for any discussion. Ultrasound has no known harmful effects on the fetus. It is therefore used when it is likely to yield information helpful in promoting the welfare of fetus or mother. It should not be used otherwise.

Remember that no matter how impressive-looking the machine, the results of investigations are only as good as the operator, the machine, and its maintenance. In many of the examinations, accuracy is not 100 per cent.

Amniocentesis

This is the process by which a needle is passed into the amniotic cavity, under local anaesthetic and with ultrasound control. It is used to obtain a sample of liquor for detailed analysis. This procedure is usually carried out at about 16 weeks in pregnancy. It may be undertaken as part of a more wide-ranging investigation when maternal blood alpha-fetoprotein levels are shown to be raised. Sometimes it is used to look for the various other substances which may indicate the presence of rare abnormalities of the fetus. It may also be performed to obtain fetal tissue cells which have come loose and are floating in the liquor. These can then be incubated so that they multiply in culture to provide a sufficient number for analysis.

This analysis can reveal a generalized fetal abnormality which has its marker in all the cells. Such a disorder is Down's syndrome (mongolism) in which the chromosomes of the nucleus are abnormal (the nucleus is the part of the cell which is responsible for determining the form and function of each cell and of the organs and body as a whole). This abnormality can be reliably demonstrated by studying the chromosomes. There are two bases for Down's syndrome. The much less common one, which is unrelated to the age of the mother, is due to a tendency inherited by the fetus from her parents. This tendency can be detected by analysing the blood of the parents or a Down's syndrome sibling. If present in the mother (the mother is a carrier), then there is a 1 in 5 chance of a baby being affected: if in the father (the father is a carrier), then there is a 1 in 50 chance of an affected baby. The other much commoner form of Down's syndrome is age-related and occurs quite unpredictably but with increasing frequency as the mother's age increases. The chances of a woman giving birth to a Down's syndrome baby increases quite quickly from about the age of 38–40 years. It occurs 1 in 1700 at the age of 20, 1 in 400 at 35 years, 1 in 200 at 37–8, 1 in 80 at 40 years, and 1 in 50 at age 45.

Amniocentesis, apart from providing this major assistance in the diagnosis of the fetal condition, has definite disadvantages which have to be taken into account when its use is being considered. Probably

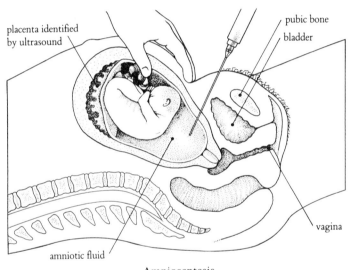

placenta identified by ultrasound

pubic bone

bladder

vagina

amniotic fluid

Amniocentesis

between 0.5 and 1.5 per cent of normal fetuses, which would not otherwise have miscarried, are aborted because the test has been carried out. In addition, about 3 per cent of babies whose mothers have had an amniocentesis have temporary breathing trouble in the newborn period, or have minor postural limb deformities which do not have long-lasting effects.

For the non-hereditary form of Down's syndrome, there is debate about the maternal age at which it is reasonable to undertake amniocentesis to exclude it. How do we determine this age? We are, of course, speaking only of those who live in a country whose laws would permit a termination and of those people whose religions and other scruples would allow them to have a termination because of an affected fetus.

There is no easy answer and it will always be arbitrary. Often the age is determined by the facilities and expertise available to carry out the investigation effectively. It could be 40 years, 38 years, or 35 years. Some doctors are prepared to administer the tests to younger women. Probably the age limit will come down as available facilities improve, efficiency of methods increases, and the likelihood of side effects diminishes. Later in pregnancy amniocentesis can help in assessing the severity of a rhesus blood group problem or in determining the maturity of the fetus's lungs if there is an indication to deliver early. An induction would be postponed if the baby was likely to develop severe breathing problems after birth.

Fetoscopy

This is another technique for examining the condition of the fetus in which, under local anaesthetic, a very small telescope can be passed through the abdominal wall into the amniotic fluid. It is essentially the same procedure as for amniocentesis as far as your wife is concerned, but the instrument has a slightly larger diameter. This is usually carried out between 15 and 20 weeks when indicated. The fetus can be inspected for abnormality. Blood samples can be taken from the umbilical cord vein, or from the fetal vessels running over the surface of the placenta. This allows analysis of the fetal blood for the purposes of diangosing many different disorders. One of these is the serious anaemia called thalassaemia, affecting people of many races but particularly those from Mediterranean and Asian countries. Current research is aiming to replace this method of diagnosing thalassaemia by a method which takes for analysis very small pieces of the placenta obtained by passing with ultrasound control a small instrument through the vagina and cervix in very early pregnancy (chorion biopsy). If this new method of earlier diagnosis proves practicable, it will enable pregnancies to be terminated much earlier when results indicate that this should be considered. This will greatly reduce the emotional and physical trauma and heartbreak associated with late terminations.

The fetoscopy method can also allow blood and other materials to be infused into the fetal circulation. Occasional very severe rhesus blood-group problems can be managed in this way.

The main problem with fetoscopy is a greater risk of stimulating the abortion of a healthy fetus (about 5 per cent). The seriousness of the possible problem to be diagnosed or treated needs therefore to be correspondingly greater than that justifying amniocentesis, if this greater risk is to be acceptable.

If any of these tests is available it is usually offered but, if you wish to, you and your wife should raise the possibility of having one. If possible, go along with your wife when any of these tests is to be discussed. You need all the advice and guidance available to help you and your wife decide whether or not to have one of these special tests performed and to agree to, say, having a termination if this is advised.

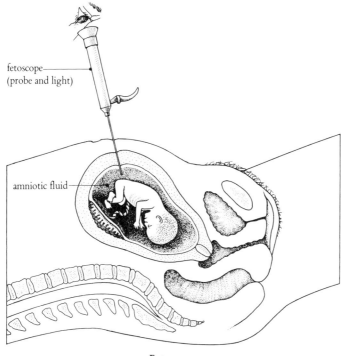

fetoscope—
(probe and light)

amniotic fluid—

Fetoscopy

Termination

If you and your wife do decide to have the pregnancy terminated, because of a fetal defect, your role in supporting and comforting your wife can be invaluable. Try to arrange to be with her during the time of the termination. Most times this will be effected by injecting a solution into the uterus (as in amniocentesis) or a tube will pass through the vagina into the uterus to carry the fluid. A solution may also be run into a vein in the arm. The termination is like a miniature labour, but can be very painful so that analgesics and extra reassurance and support are required. You and your wife may or may not wish to see the fetus—at least you will probably want to know the sex.

The following weeks and months may not be easy. Take any opportunity offered to ask questions either about the present or the future.

You and your wife will have suffered a loss which may not be fully understood by your friends. Afterwards, you will feel mixtures of relief, guilt, sadness, and anxiety. Although you have had only a partial pregnancy, the loss of your expected baby in this way can be depressing and difficult. You will need to talk over your feelings and give way to and share your grief that the pregnancy has ended so sadly. Read the section on stillbirth and on abortion. Your situation is not dissimilar.

8 *Antenatal classes*

The types of classes

In many districts a wide variety of antenatal classes is offered. These may be in your hospital or in hospital-associated clinics in the community. There may be others run by your GP or health centre, by local midwives or health visitors or by the National Childbirth Trust. Most are free but you pay for NCT classes and certain other privately run classes.

Husbands are usually welcome and encouraged to take part in classes, but as many classes are in the daytime it may not be easy for you to spare the time from other commitments. There are usually at least one or two evening classes arranged specifically for husband participation.

Classes vary from district to district and within the same district in their aims and scope and in the abilities and experience of the people leading them. How do you and your wife decide which ones to attend? First I suggest you re-read Chapter 1 and paragraph 2 on p. 20 if this is your first baby, or Chapter 1 and 2 if this is your second or subsequent baby. This will give you background information. You should then talk to your GP, hospital or community midwives, hospital medical staff and your friends—especially those who have had a *first* baby recently. Fortunately there is considerable co-ordination of antenatal classes in most districts and this will increase.

For those delivering in a hospital, classes should ideally be arranged at the hospital or under its auspices as an integral part of the antenatal care programme. There are many advantages in attending your hospital classes. You and your wife learn how care and support are arranged where she will actually deliver and she becomes familiar with the atmosphere and ideas of her hospital and some of the staff as well as with mothers who have experienced this hospital. However, although they are improving not all hospitals run comprehensive classes which take account of and discuss feelings and emotions as well as the realities of the mechanics and procedures of labour and its management. There should be discussion of baby care and baby feeding and the way you and your wife are likely to feel.

Try to find out about the hospital classes and if they are not good you may get some additional help from outside classes. Unfortunately the above reservations also apply to many outside classes. If you and your wife attend several sets of classes with a view to deciding what is best for you, you are in danger of becoming confused. There is no magic formula which will ensure an easy labour. You and your wife should by the end of pregnancy be in a position to approach labour with open minds, clear general principles and a feeling of trust and confidence in your attendants. If the advice you are being given at classes is different from what seems to be hospital practice then try to discuss the differences with one of the more senior doctors or midwives.

Beware of being too influenced by the medical literature put forward by non-medical people. There is a wide spectrum of opinion in the literature but an understandable tendency for people to quote only articles or research which tends to back up their already formulated opinions. In other words it is unusual to be presented with a balanced and non-partisan view.

It is useful to have recorded in a patient's case notes some points about her wishes. These need to have been discussed. For example, 'would rather not be induced'. Whether or not this is done depends on developments and the circumstances at the time. 'Would like epidural when labour is established.' This again isn't a rigid instruction but an orientation, as the patient may change her mind by the time labour starts. 'Would prefer to have a general anaesthetic if needs caesarean.' This means that she is likely to have the general anaesthetic unless she changes her mind. We know that most people would like, if possible, to have a normal labour and delivery but we also know that a few would very much prefer to deliver by caesarean. That some would like as little analgesia as possible, while others would like as little pain as possible. Any of these preferences is subject to variation at the time. How much should be written down and how much notice should be taken of it in the changed circumstances of labour is difficult to decide.

This is why I would say that to have a rigid plan or rigid attitudes and expectations about the many aspects of labour and delivery is not practicable or in your wife's interests. Any plan for labour should be formulated only after learning a reasonable amount about the subject from informed sources. Provided it acts merely as a general guide to be modified with time and provided it relates to aspects where some degree of choice is not against your wife's interests, then it is reasonable to have a plan and expressed preferences. In practice too much detail is likely to be counter-productive. If there is a rigid plan for labour or over-optimistic or inflexible expectations than if significant deviations become necessary they may not be easily accommodated by your wife. The plan and expectations can be responsible for unnecessary disappointment and disillusionment which may have long-lasting consequences.

The content of the classes

A principal function of antenatal classes is to give extra time for you and your wife to learn more about pregnancy, labour, delivery, and child care. They are a useful forum for discussion and a place to meet and share your thoughts, feelings, and anxieties with others. You are encouraged to ask questions, talk about your various anxieties, and to realize that others have similar problems. You are informed about the experiences which lie ahead. We aim to tell you, and particularly your wife, how she will be affected, what she will feel, what doubts and anxieties may arise in the future, what problems she may meet.

We aim to teach you and your wife what she can do to help herself and how you can help. The breathing and relaxation techniques which may be employed are an important part of instruction. You should learn about them from your wife if you are unable to attend the classes, and help her to practise by doing the breathing and relaxation with her. It is often helpful if you produce pain intermittently by giving her a 'Chinese burn' on the arm, which slowly reaches a peak and slowly wears off as would a contraction—but don't be too unkind! She concentrates on trying to relax and on her breathing with this pain as she will do with the pain of first stage contractions.

We also talk about how we, the attendants, will help with encouragement, explanation, careful observation, and examinations which may include special techniques such as electronic monitoring where appropriate. Analgesics are discussed rather as I do elsewhere in this book. Various aspects of baby feeding and baby care and of your development as a family, with the attendant problems and triumphs, are introduced and discussed.

The personnel involved and their roles in the classes depend on the practical experience and teaching abilities of those available. As much as possible, staff who are and will be caring for your wife take part. In our classes there are usually obstetric physiotherapists, obstetricians, midwives from the labour and postnatal wards, paediatricians, district midwives and health visitors, and last, but by no means least, recently delivered first-time mothers with their babies, from the adjacent wards. These women who have each had a recent experience of a normal or forceps delivery or caesarean section are much more likely to give realistic accounts. It is so easy for memories to become modified by the passage of time, by rationalizations, and by self-justifications. For 'preparation for labour' classes we much prefer to have those expecting first babies separate from those expecting second or later babies. As the two types of labour are so different it is confusing to have both presented to a mixed group. It is difficult enough to grasp all the points about one or the other. Multigravid women have different requirements which are not met when they share a class with first-time mothers (*see* Chapter 2).

A tour around the hospital is a valuable contribution to preparation for the hospital stay and you should try to go with your wife to see the admission and delivery rooms, the post-delivery accommodation and the arrangements for the baby. This is best attended after most of the antenatal classes. It is another good chance to ask questions and to find out something of the strange ways of hospitals—most enjoy this introduction and find the hospital less forbidding than expected.

Many specific techniques and regimes of breathing control, relax-ation, etc. have been advocated. Most of them have good points as well as bad. Our current teaching takes account of these as well as adding original ideas based largely on what we have observed and learned from patients. Above all, any regime likely to be useful for the stresses of strong labour must be simple. It must also be well known and under-stood by the labour ward attendants so that they can give useful encouragement and support. This is a principal reason for advocating antenatal classes within the hospital complex. You both need to be well informed in advance about the techniques, management, and attitudes of those who will guide you in labour. You also need to develop some understanding of how you and your wife are likely to think, feel, and react at the time.

Although, during labour, progress, problems, and management will be explained and discussed as they are met, there is often not time to go into much detail. Your wife may not be in a sufficiently calm state to either take in explanations or to ask questions and react reasonably to ideas which are new to her. You, also, can at times be fraught, anxious, and worried. To see someone you care for in pain and distraught is distressing for you and removes much of the objectivity which you had probably assumed you would have.

For instance, there is no aspect of labour where these considerations are more pertinent than in the provision of pain relief. There can be scope for choice of whether or not to have pain-relieving drugs in a particular situation. There can also be choice in the type of pain relief. Without quite extensive prior consideration of the advantages and disadvantages of the various options, it is often quite unrealistic to attempt to exercise this choice at the time. It is much better to have some prior understanding so that the considerations are at least semi-familiar.

You and your wife need to develop an attitude or philosophy in approaching labour. It is reasonable to hope, or perhaps expect, that your wife will have the type of labour and delivery with which she will be able to cope, with minimal analgesia provided she is given sympathetic guidance and management. However, it is essential to keep a very open mind. The range of the possible experiences which lie ahead is very wide. Nevertheless, when the time comes, it is essential to be able to accept as normal under the circumstances whatever course labour and delivery happen to take in your wifes individual case.

If the course of events necessitates an induction of labour, intravenous drip, epidural anaesthetic and then forceps, or caesarean delivery, when you had been hoping for everything to be natural, then the way you both view this apparent misfortune is very important. Rather than disappointment and resentment and an attempt to blame others for what you and your wife see as failure, you could be very grateful and happy that there are these safe and efficient escapes from potentially disastrous situations. I would not suggest that everything that happens, or that all the advice that you get is always the very best, but I have seen women and their husbands unnecessarily bitterly disappointed when everything has, under the circumstances, gone as well as possible. This is the very negative result of over-optimistic expectations.

The expectations which you and your wife have for labour and delivery are major factors in determining the quality of your experience, and the way you view it at the time and in retrospect. When your wife is in the midst of labour, it is important to realize that trust and confidence in those guiding you and your wife are very important ingredients contributing to the success of the enterprise. Of course, blind and complete trust may also be inappropriate. You should be sure not to be put into the position of being used, during labour, as pawns by individuals or groups trying to bring about an alteration in obstetric practice. You can play an important part antenatally by going with your wife to antenatal visits or antenatal classes. You are then likely to have a better understanding of the philosophy behind management attitudes you may meet in labour. For instance it can be helpful to have discussed beforehand the reasons for making an episiotomy as there is little time to consider them at delivery. You can sometimes remind the staff about your wife's preferences for or against the different types of analgesia, but remember that your wife's wishes can change radically in the real situation.

Breathing techniques

A general principle with breathing techniques is to try to keep in check the total amount of breathing. Women in labour tend to over-breathe anyway, and then tend to accentuate this if they are consciously using a breathing technique. Over-breathing doesn't matter too much, but it can give your wife 'pins and needles' and make her hands go into spasm. (This is relieved by breathing in and out of a paperbag to build carbon dioxide concentrations up to normal.) Perhaps more important is that, with excessive over-breathing, arteries can go into spasm and the blood flow to the placenta can be reduced. This is very unlikely. The emphasis is therefore on breathing more shallowly as the rate of breathing inevitably increases as labour advances.

Earlier in the first stage of labour we like women to be aware of, and consciously control, their breathing at about the same rate and rhythm as normal breathing. When labour is well advanced your wife will find it easier to breathe through her mouth even though this tends to make her mouth dry. As contractions become stronger, she will automatically tend to breathe more quickly and therefore she should aim to breathe more shallowly. When contractions are long and strong in late first stage, many women find it easier to concentrate on shallow, panting breathing and often to group these in threes. This breaks a contraction into short segments, and makes it easier to deal with. These difficult contractions may be the pattern of a first labour for only one to two but sometimes for many hours. If, as often happens, the feeling of wanting to bear down develops towards the end of the first stage, before the cervix is fully open, it is easier for your wife to resist this feeling if she breathes shallowly in threes. You can do this breathing with her or count her breaths for her, or squeeze or tap her hand in time with the breathing. Remember that your wife is probably not ready to bear down until she cannot stop herself from doing so. (Even then, subsequent examination sometimes shows that the cervix is not fully open.) It is very easy for you, sitting through a longish labour, to try and will her into the second stage. To say, 'Are you sure you are not ready to push, dear?' is quite the wrong approach. By doing so, you can easily precipitate her into pushing before the cervix is fully open. The emphasis must always be on not pushing until the urge is irresistible.

During the first stage your wife deals with one contraction at a time and you can remind her that each one is one closer to her goal—in the same way that each step taken by a mountaineer brings him one step closer to the summit.

During contractions, your wife concentrates on her breathing, tries to relax, keeps her eyes open—usually fixing on a spot—and holds your

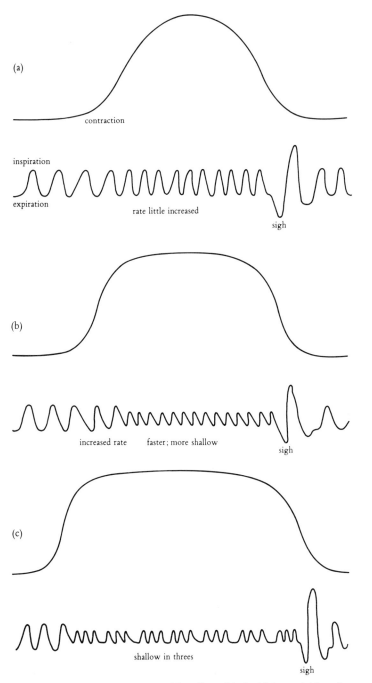

(a)

contraction

inspiration

expiration

rate little increased

sigh

(b)

increased rate   faster; more shallow

sigh

(c)

shallow in threes

sigh

Breathing in labour: (a) early first stage; (b) well-established labour; (c) late first stage

hand. She may hold your hand so tightly that *you* have to start breathing, so insist that her nails are cut when labour is due. Whether or not she actually relaxes is not important—it is rather that she tries to—when contractions are strong, holding on to your hand, concentrating on the breathing, and fixing on a spot may be all she can manage. Encourage her to let go and relax between contractions. As a contraction comes, say to your wife, '*breathe*, open your eyes, look at your spot and be loose', but emphasize '*breathe*'. If there is a monitor you can tell her when the contraction has reached its peak. Otherwise you will judge the peak by timing the length of a contraction. If she falters in the breathing, encourage her by saying, 'Keep breathing. You are doing very well.' She is often thinking, 'Will it *ever* pass the peak and go away again?' Say, 'It is starting to go down.' Sometimes it is helpful if, during a contraction, you rub her lower back firmly in time with the breathing. Sometimes just firm pressure is helpful, but some women find this irritating. Various modifications of lying on the side with the help of pillows can be comfortable, as well as favourable for the progress of labour. Early in labour, your wife may prefer to be active and to walk about, pausing to lean on you or on a chair or table during a contraction. She may sit in a chair for a time. Usually when labour is advancing strongly she prefers to lie on her side, or in a propped-up position, or in a chair. The only posture we don't like is lying flat on the back, as this may slow the

Leaning on you or against the wall during contractions as labour becomes established

progress of labour. The best posture to adopt, within the limits imposed by her labour and its management, is the one she finds that suits her. More important than a particular posture is to be able to change her posture.

Once second stage is established and your wife has an overwhelming desire to push down, or when the baby's head is found to be low in the pelvis, she bears down with contractions. She is usually most comfortable in a supported semi-sitting (supported squatting) position, and this we have found the most efficient posture for second stage. A mirror can be helpful to show her the first glimpses of the baby's head. For the actual delivery your wife is slid down a little to make for easier delivery of the baby's shoulders. These postures allow good face-to-face contact between the patient and the midwife or doctor. This further ensures

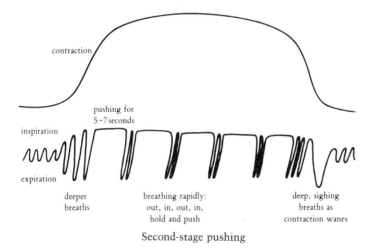

Second-stage pushing

good control which is so essential. Your wife can watch your baby emerge and the midwife can easily control the emergence and see the perineum to make an episiotomy safely and efficiently if it is necessary.

There has been interest recently in returning to the use of some form of birth chair. However, development has not so far reached a stage where models available at a reasonable price are sufficently comfortable and free of problems. Research and development are proceeding.

The technique of efficient bearing down is a skill which your wife will have been taught. As a contraction starts, she spreads her legs well apart and relaxes them, holding them to fix her trunk. She takes a moderate breath through her mouth, closes it, and holds the breath as she pushes with all her strength towards the anus. At the same time, she has the feeling that she is letting go and is opening out 'down below'. She must go with the contraction and resist any tendency to hold back or tighten the muscles around the vaginal opening.

After only five to six seconds, she quickly exhales, quickly breathes in and then out again, and then holds the next breath and pushes for another five to six seconds, and repeats this as long as the urge to push continues. Her legs should remain relaxed and preferably supported by attendants. She should not push into her legs or push her feet against anything, as this tends to tighten the muscles of the pelvic floor and to slow the baby's progress by providing extra resistance. During pushing your wife should not hold any breath longer than seven seconds, as by so doing the oxygen in her blood is lessened and less blood flows to the placenta. The oxygen in the lungs is used up and the back-pressure caused by the tension in the chest stops blood flowing to the heart. There is then less blood to be pumped to the placenta.

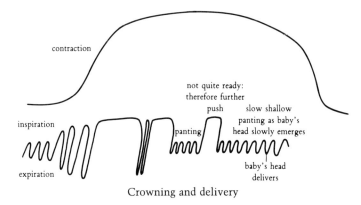

Crowning and delivery

As the baby is delivering she stops pushing and pants slowly and shallowly through her mouth. This ensures a slow delivery with only the force of the uterine contraction acting. If the baby doesn't quite deliver and the contraction wears off then we ask your wife to bear down between contractions to effect the delivery, i.e. *either* the uterus *or* she can push to effect the delivery but not both together as the baby may emerge too quickly.

9 When to go into hospital or to call the midwife for home delivery

The general principle is that your wife should be observed by a professional during the whole of labour. Labour is the time when unexpected problems are more likely to arise, no matter how normal the pregnancy has been up to that time.

I explained in the section on the placenta how the baby's oxygen supply may be in jeopardy once labour starts.

The onset of contractions

The time to make the move is therefore when it seems likely that labour is under way. For those having a first baby, I advise that this is when contractions have been coming on average once in 10 minutes for three hours. If the average time between contractions is less than 10 minutes then go sooner. In rapid labours contractions can be close together from the start. For a second or later baby, it is when your wife thinks she is in labour—usually with contractions coming at 10- to 15-minute intervals, or more often. Women having a second baby usually know the difference between labour contractions and false labour, but they can be wrong. It is not uncommon, even when labour is establishing, for contractions virtually to stop on arriving at hospital, but they usually start again within 30 minutes or so. If you are in doubt, you can always ring your hospital midwife or your doctor and get their advice. Most hospitals like you to telephone before going in so that they can find your records and be ready to receive you.

Some women have a bout of diarrhoea at the onset of labour which may continue for some time as labour advances. Don't make the mistake of ascribing lower abdominal colicky pain and diarrhoea to an attack of stomach 'flu or to having eaten an unusual food.

Don't forget premature labour—labour can start at any time. When the baby is old enough to have a chance of surviving we speak of premature labour, rather than threatened abortion which is the appropriate term for labour earlier in the pregnancy. Babies can survive and do well when born as early as 16 weeks before the expected date. If your wife starts having recurring lower abdominal period-like cramps or if she has low back-ache which may be continuous, with superimposed recurring stronger pain, you should seek help without delay. If the problem is threatened premature labour, this can often be stopped by dripping a solution of one of several substances into a vein. This may well enable the baby to stay inside until he is more mature and can be born with a much better chance of survival.

The 'show'

The 'show' is much talked about. It is the loss of blood-stained jelly-like material (blood-stained mucus) from the vagina. This mucus plug usually fills the cervix during pregnancy, and is part of the mechanism for preventing bacteria from gaining access to the membranes or fetus. As the cervix starts to take up and dilate, the mucus is released and is stained by the slight blood loss coming from the movement of the membranes. The 'show' is therefore related to the onset of labour. It may not be released until labour has been in progress for some time, or may be released a few days before labour becomes established. It does not call for action and is not in itself a reason for coming to hospital.

In sharp contrast is the loss of even one drop of fluid blood. This, or any more extensive bleeding, should be reported without delay. It may be a warning that the placenta is attached low down on the uterus, close to the cervix (*see* antepartum haemorrhage).

Rupture of the membranes

Loss of liquor is another indication for hospitalization. Although the waters usually break in the first stage of labour, they may break at its onset or even weeks before labour is due. The loss of liquor may be either as a sudden rush or as a slow trickle.

With a sudden rush, the baby's cord may be swept down in front of her head (prolapse of the cord) and could then be compressed during contractions should labour start. You would not see it unless it happened to be right outside. The baby can tolerate this for a time but would need delivering by caesarean section as soon as possible. The waters may break conveniently at home, but your wife could be out shopping—she may notice a slight pop and then have her legs drenched with fluid. She will no doubt step aside quickly and look around to see who could be responsible for the puddle. There is no need for her to panic, but she should arrange to go to the hospital by car, taxi or ambulance without delay. It is probably not a good idea to explain the details to the taxi-driver as he may prefer to keep his taxi liquor free, but a pad in the handbag could be a wise late pregnancy precaution. In hospital, she will be examined to exclude prolapse of the cord (coming down into the vagina or at least below the head). If no cord is found, she would still be kept in hospital because of the chance of infection developing. Prolapse of the cord occurs about once in 500 deliveries.

If there is a continuous trickle of fluid or a constant dampness of the underclothes, there may be a small leak. This can be confused with loss of urine which is not uncommon in late pregnancy (The urine loss is intermittent and usually in small amounts with laughing, sneezing, or coughing.) The small amount of discharge normally present in the vagina is of an acid quality—the acid is important in inhibiting growth of potentially harmful bacteria. The escaping liquor is alkaline and interferes with this defence mechanism. Any woman with leaking liquor should therefore be undergoing observation in hospital to detect and treat any early suspicion of intra-uterine infection. Encourage your wife to report any suspicion of such a leak. If the pregnancy is within about four weeks of the expected date and labour doesn't start within a few hours, an oxytocin drip would be used to stimulate the uterus. The baby is of an age when she is safer delivered than running the small risk of infection within the uterus.

Go to hospital when you have any one of the following signs:

1 **Water breaks or leaks**

2 **Bleeding late in pregnancy**

3 **Contractions coming**
   (a) **for a *first* baby at intervals averaging one every 10 minutes for 2–3 hours *or* if contractions are coming more frequently**
   (b) **for a *second* or later baby when your wife thinks she may be in labour**

10 *Unexpected delivery*

How will you cope if faced with unexpected delivery? Well, first of all there should be very little chance of this happening if you follow the rules set out for going to hospital, or for calling the midwife if you are planning a home delivery. The idea is to have ample time for the journey and to be settled in, so that labour is supervised. Women are still, unfortunately in my opinion, sometimes advised to stay at home as long as possible. Those who haven't much experience of labour and its potential problems and therefore are not qualified to advise are sometimes known to say, 'It's nicer to spend most of the first stage at home and to go into hospital for only the last few hours of labour'. This is the advice that can result in deliveries in the worst possible circumstances—perhaps in a car or on a street corner, with no facilities and no expert assistance—dangerous for both mothers and babies, and decidedly nerve-racking for fathers! There are other dangers—failure to detect the baby becoming dangerously short of oxygen, the placenta separating in first stage and haemorrhaging, the baby's cord coming down—to mention just a few.

Remember the rules given on p. 77 for when to go into hospital. Very unusually, contractions with a first labour come so close together from the start as to seem continuous—an indication for rapid action. If in doubt, telephone the labour ward, your midwife, or doctor. With second or later babies, you make the move as soon as your wife thinks she is in labour. This is usually with intervals averaging 10–15 minutes, but with a feeling that makes her think this is true labour rather than a false labour. If in doubt, ring the labour ward, but don't be easily put off from going in. Your wife is much more likely to be right in diagnosing the onset of labour than anyone on the end of a telephone—especially if it's 6–8 a.m. (the end of the night shift). You are most likely to go wrong when the first labour has been long, dreary and painful. After such a labour, your wife is likely to say, 'The last thing I'm going to do is to go into hospital early this time.' You are tempted to expect the same pattern, and wait about at home until strong contractions force you both to make a move. With a second labour, this may be only a short time before a short second stage. As you move out the door, your wife is likely to grab you and say, 'Do something—the baby's coming!' and you are expected to spring into action!

Preferably move back inside, get your wife to lie down and remove at least any nether garments. If the baby is coming you may be able to steady his advance by placing a hand gently on his head and firmly telling your wife to stop pushing and to pant. The chances are that he will surprise you by his speed of entry as your wife ignores your command. It is quite a good idea to hold him slightly head down until he

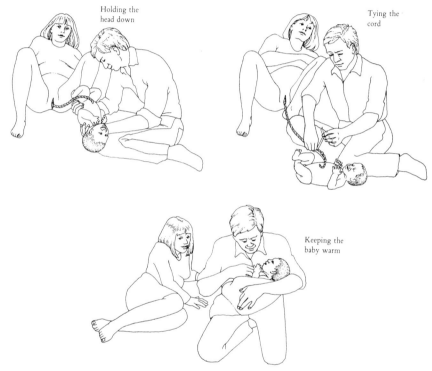

A baby often has a bluish tinge when he is first delivered. He soon cries and starts breathing. (After an illustration in *Having a baby* McDonald Publications.)

cries, just so that he doesn't inhale too much fluid. He is likely to be in good condition and to breathe and cry without delay. He is wet and will soon get cold, so quickly wrap him well—preferably after quickly drying him off. Put him on your wife's tummy to help keep him warm.

Should you tie the cord? Not essential, but preferable. It must, however, be tied only with something wide like 1 cm tape—don't use a piece of string or whip off a shoe lace, as this is likely to cut through the friable cord. If you are a DIY action-man type, you may like to keep a piece of tape in your pocket. At least it'll remind you not to hang about at home as you may then need to use it! You don't need to cut the cord, but if you feel you must, then first tie it in three places—5 cm, 7.5 cm, and 9 cm from the baby's abdominal wall. You cut between the second and third ties from the baby's abdomen. It doesn't matter if the placental end bleeds, but the baby's end must not. If your knots are not good enough and it does bleed, your wife can pinch it with her fingers while you call the ambulance. If the placenta delivers, you should tie the cord and then wrap the placenta with the baby—they are used to being with each other and don't mind. Most important—don't let the baby get cold.

Ask someone to call an ambulance or call it yourself, and continue your interrupted dash to hospital where they will attend to any stitching and even make you a congratulatory cup of tea!

11 Labour

This is the process at the end of pregnancy which results in the expulsion (birth) of the baby and the expulsion of the placenta. The birth of the baby is divided into the first and second stages, and the delivery of the placenta is the third stage.

During the first stage, the cervix is pulled open by the action of the muscle of the uterus. When it is sufficiently open to allow the greatest diameter of the baby's head or breech to pass out, the cervix is fully dilated, and the first stage is complete, marking the start of the second stage.

At the start of labour in women having a first baby, the cervix is often long and its central canal is only half to one centimetre in diameter. The action of the uterus in pulling on the cervix is first to reduce the length of the cervix—the process called taking up (or effacement) of the cervix. It is then gradually stretched open—the process of dilatation of the cervix. Dilatation is described in terms of the diameter of the circular opening of the cervix. Progress is often very slow and almost imperceptible until the cervix reaches some three centimetres in diameter. From then onwards dilatation is increased by about one centimetre per hour until full dilatation at about 10 centimetres. In many women, much of the work of labour is effected by the uterine contractions of late pregnancy. Occasionally, by the start of labour, the cervix is already completely taken up, thinned, and as much as three centimetres dilated. Not surprisingly, labour is usually much shorter in these women. The length of the first stage of labour varies quite widely but averages something like 10–15 hours. It can, with first babies (rarely) be as short as one to two hours or may be very much longer than 15 hours, even when mild early contractions are not counted. In many women, the uterus stops and starts and proceeds slowly for quite some time before it finally settles down to what we call established labour.

Some of the short labours can be violent and very painful, and it may be very difficult to ensure adequate pain relief, partly because it is not appreciated except in retrospect how rapidly progressive the labour is. Contractions may start at two to three-minute intervals and soon seem to be continuous. After these labours the woman can feel that she has not coped well and can be rather shocked by it all.

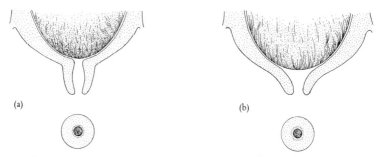

(a) Cervix long and closed at the onset of labour (an 'unripe' cervix); (b) Cervix partly taken up, but still closed

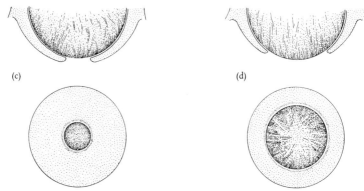

(c) The cervix has been pulled open (dilated) further; (d) Cervix completely taken up and starting to open (2–3 cm dilated). Some women start labour like this and, as a result, have a shorter labour. The cervix is 'ripe' and labour would progress well if there were an induction

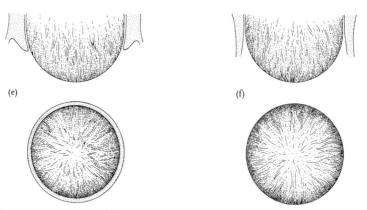

(e) The cervix is nearly open (a rim of the cervix—often called the anterior lip—is present); (f) The cervix is fully open—fully dilated—end of first stage and beginning of second stage

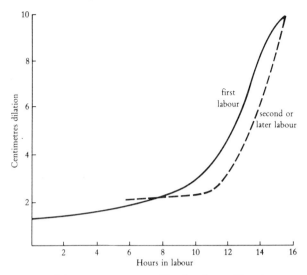

First labours progress steadily after 3 cm dilatation

How will you recognize the onset of labour?

In practice, this is often difficult and whether or not you have the real thing or a false alarm becomes clearer with the passage of time. As mentioned earlier, the contractions of early labour may not be different from those of pregnancy. The characteristic of labour is that the contractions become progressively stronger, more painful, closer together, and they keep going. They often recur at fairly regular intervals. However, they may remain irregular in strength and frequency throughout labour and for some lucky women are not particularly painful. When contractions start, they usually last 20–40 seconds and they are usually 10–15 minutes apart, and then become more frequent. When labour is well established and advancing, contractions last $1–1\frac{1}{2}$ minutes and the intervals between the start of one contraction and the next are $2\frac{1}{2}$–3 minutes. Sometimes contractions are, right from the start of labour, so strong and frequent that they seem to be almost continuous. As a general guide, your wife is probably in labour if, for 2–3 hours, the intervals between contractions average 10 minutes, or if the contractions are coming more often.

Contractions are felt as a recurring tight gripping ache low in the abdomen or in the lower back, or both. Some women feel they are getting a tummy upset like gastro-enteritis and they may have diarrhoea. Sometimes contractions seem to start in the back and move to the front, and sometimes the other way round. Many describe them as like period pains. They may be felt in the front of the thighs down to the knees. The severity of the pain of contractions varies widely from one individual to another. In general, the pain increases as labour progresses, but it can be severe from the start.

Some women feel nausea during contractions or just at the peaks of strong ones once they are advancing in strong labour. Occasional women then wretch or vomit with these contractions. An antivomiting drug can be given if this problem persists.

The baby's progress

As far as the baby's progress is concerned, I will describe the usual situation in which the baby is coming head first with his head well tucked up. This keeps his chin down on his chest. The back of the head is pointing to the side or partly towards the front. In this posture, the back of the top of the baby's head is coming first. This is as viewed from below, looking up into the vagina. The shape of the widest area of the head is circular with a diameter of about 9.5 cm. This is also the widest area of the whole baby, as the size of the head is not exceeded by any succeeding part of the baby. When the shoulders come, they are well hunched-up and are therefore no bigger than the head. As the cervix is pulled open, the top of the baby's head bulges progressively further through it, and in the process his head moves very slowly down the upper part of the birth canal. In the later part of the first stage and in the second stage, the baby's head also rotates so that by the time he is born, the back of his head is pointing directly towards the front.

During the first stage of labour, your wife's role is almost entirely passive as far as the process of labour is concerned. The only thing she can do which may shorten the first stage is to adopt a posture other than lying flat on her back. Being up and about in early labour, or sitting in a chair or propped up in bed, or lying on her side, are all the preferred postures. Otherwise, her role and concern is for her own welfare—coping with pain and anxiety, and waiting for the second stage. We will discuss the important principles of this coping process later. There is no evidence that relaxation, for instance, has any direct effect on the way the uterus works, but it is part of the coping process for the mother.

The second stage of labour is quite different in that your wife now bears down actively with contractions to assist the uterus. As the cervix is now fully open, the maximum diameter of the baby's head passes through into the upper vagina. The action of the uterine contractions is to squeeze the baby on to the floor of your wife's pelvis where the lower part of the birth canal is formed by the bulging soft tissues. They direct the baby forwards to the vaginal opening.

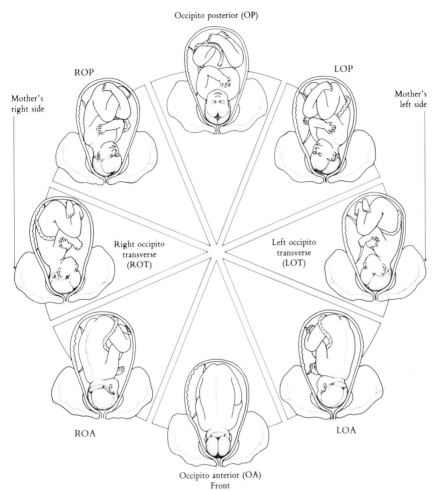

Different directions in which the baby's head points (positions) at the start of labour

The birth canal is the passage through which the baby passes from the lower abdomen to reach the outside. The stretched lower uterus and vagina are pressed against it as the baby's head fills them on its way down. The uterus and vagina form the inner lining of this birth canal. It is shaped like a curved pipe. The upper part is formed by the bones of your wife's pelvis and the lower part by the bulging stretched floor of the pelvis (the soft tissues filling in the lower opening of the bony pelvis). In the accompanying diagram, the birth canal is seen from the side as it would be if a woman were lying on her back. The front wall of the bony pelvis is the relatively short pubic bone, while the back wall is the hollow curved front wall of the sacral bone (the lower end of the vertebral column). Note the backward curve of the sacrum which is largely responsible for the prominence of a woman's bottom. The coccyx (or tailbone) at the bottom end of the sacrum is hinged and moves out of the way as the baby's head passes. It is of no importance in birth. The space between the lower ends of the bones is normally filled in by soft tissues which are for the most part muscular. During pregnancy, these tissues become progressively softened and take on an elastic quality. Just as a small rubber band can be stretched to almost any size, so these soft tissues at the opening of the birth canal can be stretched with the passage of the baby.

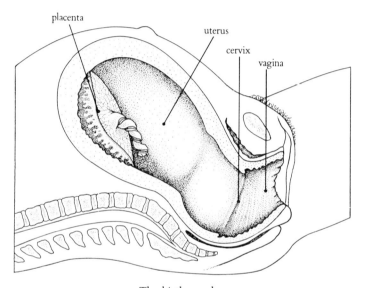

The birth canal

Note the direction in which the baby passes through the birth canal. This is at first downwards and backwards but is gradually changed by the shape of the birth canal, so that he emerges moving a little downwards but mostly forwards. When your wife bears down with the muscles of her diaphragm and abdominal wall in the second stage of labour, she can push only in a straight line. She pushes backwards and downwards along the axis of her bony pelvis. She relies on the shape of the birth canal to direct the baby forwards. She cannot push around the corner towards the vaginal opening, but concentrates on pushing backwards behind the anus, more towards the coccyx as though passing a bowel motion.

You can see from the diagrams that the baby's head takes up all the available room, so that other structures such as the lower bowel and the

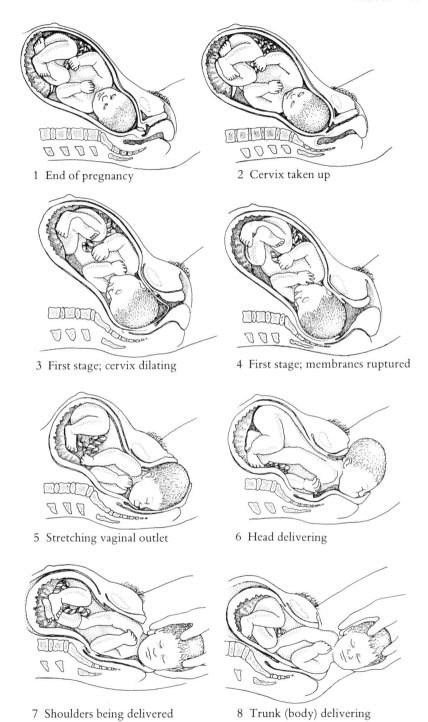

1 End of pregnancy

2 Cervix taken up

3 First stage; cervix dilating

4 First stage; membranes ruptured

5 Stretching vaginal outlet

6 Head delivering

7 Shoulders being delivered

8 Trunk (body) delivering

urethra (water pipe) get quite squashed but, fortunately, are not damaged in normal labour.

The baby's role in labour is passive. She is moved down a few extra millimetres with each contraction and recedes again as the uterus relaxes between contractions. Sometimes, especially with a first baby, this progress is very slow. This is usual and normal when there is just a glimpse of the baby's scalp at the vulva before the head has come down enough to part the large muscles of the pelvic floor. Once the sideways diameter of the head at the vulva has reached three centimetres, progress should be rapid. If it isn't, this is one of the indications for episiotomy. Once the baby's head has delivered, the shoulders follow after a brief interval and the rest of the baby very quickly. This completes the second stage of labour.

How does a woman know that the second stage has started?

The characteristic feature of the second stage from her point of view is that she has an irresistible desire to bear down as the uterus contracts. Many women start to get this feeling well before the end of the first stage and with others it does not become more than weak and indefinite even when the second stage is well advanced. We advise women to try to continue superficial breathing during contractions until they cannot stop themselves from pushing. Even then, if the baby does not appear at the vaginal opening after four or five contractions, the midwife or doctor does a vaginal examination to check that the cervix is fully open. If it is not, the woman lies on her side and will be given extra encouragement not to push and perhaps gas and oxygen, or some other form of sedation, if needed.

Your baby will not be harmed if your wife pushes too soon, but sometimes the pushing can cause the cervix to swell, leading to some delay. At very least, your wife will be unrewardingly exhausting herself. Inappropriate compulsion to push down too soon is sometimes relieved only with an epidural. You must be careful not to precipitate this early pushing. It is very easy when you are sitting with your wife as the hours tick by to say, 'Are you sure you don't want to push?' 'It must be nearly time to start pushing', etc. You must be patient, and be guided by the general principle that almost all women tend to push too soon.

Alternatively, the doctor or midwife may diagnose second stage either by looking and finding the head at the vulva in the lower birth canal or by doing a vaginal examination and finding the cervix fully dilated, i.e. no cervix to feel anywhere around the edge of the head.

What does your wife feel during the second stage of labour?

Usually at the end of first stage there is, with each uterine contraction, this gradually developing feeling of wanting to push down with the contraction so that, by the time she is definitely in second stage, this desire to push down has become irresistible. In some women, the compulsion to bear down comes fairly suddenly and can sometimes be alarmingly positive. In others, it can be almost completely absent, and bearing down is established only by positive effort after second stage has been unequivocally established by finding on examination that the cervix is fully dilated and the head low down in the pelvis. Contractions are sometimes quite weak in early second stage and usually vary quite markedly in strength. Provided your wife is definitely in second stage and provided she has the confidence to really let herself go and bear down as she has been taught, then lower abdominal pain, such as that of the first stage, is not usually a feature of second stage. Bearing down seems to relieve it. However, when she draws breath between pushes or stops pushing before the end of a contraction, lower abdominal pain is inclined to return. Some women find that the tummy muscles begin to ache after pushing strongly for a time (unaccustomed exercise!).

During active bearing down, the following sensations are experienced with contractions. (They disappear as the baby's head slips back up the birth canal between contractions.) At first, there is a feeling that the bowels are going to move as the baby's head squashes the lower bowel. This gradually changes to a certainty that she is going to pass a large grapefruit or football through the anus. Next, as the soft tissues start to bulge down, the part between the vagina and the anus (we call it the perineum) starts to stretch very slowly at first. Many women have the impression that they are pushing against a brick wall and getting nowhere. The head may for a time advance only two or three millimetres in three or four contractions. A stretching feeling gradually develops and increases with subsequent contractions until she feels as though she is going to split in two—a bursting feeling. This is a stage when real doubts start to creep in. Your wife, not surprisingly, has a tendency to hold back and not to go with the contractions—she may even have the feeling that it's time to call the whole thing off and go home! However, provided she knows that these are the normal and expected sensations and reactions, and provded she is encouraged and is confident in the midwife or doctor helping her with labour, she will find that the more she lets herself go and the harder she pushes, the better and more confident she will feel.

As the opening of the vagina stretches, the sensation is more one of burning, which may change to a numbness at full stretch. Not uncommonly with first babies, this final stretching is rather defective in that either the tissues are too resistant and stretch too slowly or the tissues show evidence of tearing. A cut (episiotomy) may be made to relieve the situation. When the need for an episiotomy seems likely local anaesthetic may be injected ahead of time. In any case it is injected into the tissues before it is made. The other main indication for an episiotomy is that the baby is showing evidence of oxygen lack and needs delivery without delay. Your wife notices the episiotomy being made as a sensation of relief—as though there is suddenly more room. It doesn't seem painful because of the local anaesthetic (for more detail on episiotomy see Chapter 20).

Whether or not the cut is necessary, the baby's head delivers next as a rather satisfying feeling of relief. As the head delivers, your wife will be asked to pant (breath slowly and shallowly through her mouth) if the uterus is contracting, or to push between contractions, so that the head delivers slowly, controlled by the midwife. The power for this delivery is thus either the uterus or your wife pushing, but usually not both together. The shoulders soon come as another lump, and then the rest of the baby slithers out fairly quickly followed by more liquor. This is again a satisfying feeling of physical relief, and release of emotional tension. Many women say the baby's actual emergence doesn't worry them.

A puzzling phenomen in some labours is recurring bouts of uncontrollable shakes—a bit like severe shivering. They can come at the end of first stage, in second stage, or following delivery. They usually pass within 10 minutes. They don't indicate that anything is wrong and I've not heard a convincing explanation.

It is important to keep the above description in perspective—many women would not agree that it is a good description of what they felt in the second stage. They may not, at the time, have been able to let themselves go, and may have tried desperately to hold back by trying to tighten their pelvic floor muscles. This, not surprisingly, may prove very painful. It is very common for women to hold back a bit without showing it, and not to commit themselves fully to relaxing the pelvic floor and pushing efficiently. The stretching of the pelvic floor and the final stretching of the vaginal opening may for some be unbearably painful, even when they are committed and co-operating fully. Feelings of panic can take over. Some women therefore have a need for analgesia—usually in the form of inhaled gas and oxygen. It is essential that a person experienced in obstetrics and who is sensitive to the woman's reactions is on hand to direct the second stage as needed. Most women need and appreciate firm but kindly instruction for second stage and delivery. Some doctors encourage women to have a pudendal nerve block to numb the opening for delivery, and those having an epidural or caudal block may wish to have it reinforced for delivery—see the section on analgesia.

The third stage of labour

This is the delivery of the placenta and membranes and is almost always nowadays assisted by the attendant. The great anxiety about the third stage is that from time to time extensive haemorrhage can occur, either before or after the delivery of the placenta. Prior to delivery of the baby, about one pint (500 ml) of blood enters and leaves the placenta each minute. When the placenta separates from the wall of the uterus the plane of cleavage is in the inner layers of the uterus just under the placenta. In this separation process, the woman's blood vessels supplying blood to the placenta are shorn across and left exposed. Excess blood loss at this stage (post-partum haemorrhage) does not normally occur because the uterine muscle has an inherent ability to contract strongly and continuously when it is no longer stretched by the presence of the baby or placenta. The muscle fibres in the uterine wall are so arranged around the blood vessels that, as they contract, they constrict these vessels and prevent blood flowing through them. This process, which usually efficiently stops haemorrhage, cannot operate if the uterus relaxes or fails to contract, as is sometimes the case.

To reinforce the natural tendency of the uterus to contact and so to minimize haemorrhage, it is routine practice to give the mother an injection of syntometrine solution into her leg muscle as the baby's shoulders are delivering. This solution is absorbed rapidly into the blood stream and reaches the uterus. Most women are not aware of this injection being given. It is not a painful injection. The injection contains two substances—one (syntocinon) to start the uterus contracting rapidly and the other (ergometrine) which is slower acting, to maintain firm contraction for at least three to four hours. If the placenta is not delivered soon after the uterus starts to contract under the influence of these drugs, it can be trapped by the uterine contraction. The attendant therefore waits three minutes, checks abdominally that the uterus is well contracted, and then delivers the placenta by pulling gently on the cord with one hand and holding up the uterus with the other hand which is placed on the lower abdomen. It often helps if your wife pushes down at the same time. The delivery of the placenta feels to her like the passage of a soft lump. It is quite a satisfying feeling, rather than one of discomfort.

The delivery of the placenta is accompanied by the loss of blood which has accumulated between the placenta and the uterine wall as the placenta separates. Anything more than 500 ml is considered an abnormally large loss. It doesn't necessarily call for any action provided it doesn't continue. Most pregnant women have a blood volume which averages nearly 6 litres instead of the usual 4.5 litres of the non-pregnant stage. There is therefore some tolerance to blood loss at the time of delivery. Occasionally the blood loss can be excessive and rapid, especially if the placenta remains in the uterus, after partially separating. Its presence prevents the uterus from contracting. Management is to give extra ergometrine, deliver the placenta (with a general anaesthetic if necessary) and give blood transfusion as required. Very rarely bleeding persists despite all efforts and hysterectomy (removal of the uterus by abdominal operation) is essential to save the woman's life.

12 The pain and distress of labour—pain relief

Most of us would like labour to proceed naturally. We also hope that it will be tolerably free of pain and distress, and that there will be no need to resort to pain-relieving drugs, tranquillizers, or special techniques such as epidural anaesthesia. You may both hope that your wife will cope using more natural methods such as breathing and relaxation techniques, massage, etc. This is a reasonable though optimistic hope, and unfortunately, experience has shown that for *first* labours such methods are usually only partly, and variably, successful and it is unusual not to need the additional aid of pethidine or epidural block although some with easier labours manage without anything or with gas and oxygen only (*see* Chapter 2 for a discussion of second labours).

Throughout the world nearly all women having *first* babies need more than gas and oxygen. After learning about analgesia many decide that, if necessary, they will have at least one injection of pethidine. The amount of pethidine a woman has depends on how difficult her labour turns out to be, how long it lasts, how much pain she has, and how much she feels able to put up with, as well as how much support and encouragement she receives from her husband and the midwife. Therefore some women will need more than one injection of pethidine and some will go on to have an epidural, sometimes after having earlier decided against having one. It is helpful for you to know this so that you don't try to push the 'no drugs' concept. You can also help your wife to come to terms with what she finds necessary to enable her to cope with the pain she is experiencing. Don't make her feel guilty or a failure for accepting pain relief. On the other hand, if in labour she says she doesn't need pain relief, then she should not be persuaded to have it. Listen to what your wife is saying *at the time*—she is the one having the baby. When Queen Victoria was faced with reluctant doctors who objected to her insistence on the new chloroform being used to ease a royal birth she is reputed to have royally clinched the argument with the words, '*We* are having the baby; *we* shall have the chloroform.' Of course, chloroform has long since been superseded.

When you and your wife talk about your forthcoming labour, your wife may charge you with making sure she doesn't have analgesics in labour. Don't take this duty too seriously, as your wife's whole attitude is likely to change when she meets the reality of the pain of her labour. She cannot plan her analgesia beforehand other than to say she would like to cope without drugs if it seems feasible at the time, or to express a preference for an epidural either electively or instead of pethidine. It would be quite wrong for you to try to prevent her having what she needs in the form of pain relief. In labour the decision to have or not to have analgesics isn't an easy judgement. It is better left mostly to your wife and the midwife or doctor. It is your wife who experiences labour and delivery, and your role is that of a supportive and involved on-looker.

I occasionally come across a husband who has become so influenced against the use of drugs in labour that he can be quite unreasonably obstructive. I think it is out of place for a husband to put unreasonable pressure on his wife in this way. However this is not a problem with most men as they find it difficult to see their loved one in pain.

Midwives have told me how annoyed they have felt on hearing something like the following exchange:

She: 'I can't cope any longer. I must have something for the pain. I want some pethidine', or 'I want an epidural.'
He: 'No, darling. You don't need it. You coped very well in the classes with your breathing.'

She is asking for pain relief but he doesn't appreciate the difference between the simulation and discussion of 'pain' in classes and the real pain of labour! In these circumstances the woman will be given the pain relief. All her husband has started is the process of making her feel guilty and a failure. A midwife was telling me how she recently noticed her patient was starting to become distressed, and asked if she would like an injection of pethidine whereupon her husband answered 'No *we* don't want pain relief!' Now we could all agree that he probably didn't need it and that he may have been voicing the opinion she held before experiencing labour. We are however much more interested to find out what she feels she needs in the labour situation.

For most women having their *first* babies, labour is a very difficult undertaking. It is associated with a varying amount of pain which is often severe and for which they usually need pain relievers, epidurals, etc. First labours are in general more difficult and painful than second labours because the uterus works less efficiently and the cervix and all the tissues below it are being stretched for the first time and are therefore more resistant. It all seems to improve with practice. I have made a study of these pain problems over some 30 years and have tested almost all the regimes which have been recommended or lauded for easier birth. Most have had good points, but claims have in general tended to be made prematurely, or exaggeratedly. I have drawn together good points from these regimes and other ideas developed from our own experience. Ideas and truths have been contributed by many people, most importantly by former patients. Our whole approach and group programme is one which has proved highly successful in our large hospital department. Our aim is to make labour and delivery at the very least an experience which women and their husbands find an acceptable start to their new life as a family. As I have already mentioned, the expectations of a woman and her husband for the experience of labour and delivery are crucial in determining how well they negotiate them.

The most positive aspect is the outcome—giving birth to or having delivered a healthy baby: this is the essence of the joy and sense of achievement. Labour in itself is very much of secondary significance as far as this joy is concerned. The positive aspect is not so much the experience of labour as the giving birth in the manner which is most appropriate in the circumstances. After the first stage of labour becomes painful it is for almost all women having a first baby anything but a marvellous experience—a time to be coped with in one way or another in anticipation of the birth. The second stage can be enjoyable, exhilerating, and satisfying for many. The earlier part of the second stage may still be difficult and uncertain until they get into the rhythm of pushing. Some woman find all of second stage difficult and painful.

When the baby's scalp is beginning to appear at the vaginal entrance, progress can be almost intolerably slow. Your wife may well say that she feels as though she is pushing against a brick wall and making no progress. A little later, as the baby's head starts to stretch the soft tissues of the floor of the pelvis, doubts and fears can flood back so that much encouragement and reassurance is required. The final stretching of the vaginal opening is very painful for some women whereas most don't find the emergence of the baby nearly as painful as the end of first stage.

The first stage requires most of our attention. The pain and distress involved are to a large degree interdependent and closely related. The approach to their management is first and foremost by educational and psycho-emotional methods. Pethidine and gas and oxygen (drugs) and medical techniques for pain relief (such as epidurals) are used as required against this background of emotional preparation and support. Labour is an emotional experience, rather than an intellectual exercise. Knowledge of the processes and mechanisms of labour can undoubtedly be reassuring, but it is by no means proof against the emotional problems which beset women in labour.

In the usually very stressful situation of labour, the reassurance and encouragement of those present are of great value, but it is the emotional reserves of the woman which are crucial. Techniques for pain relief, such as epidural, may sometimes be needed as much because of emotional difficulty as for pain. On the other hand, no matter what the emotional strengths of the woman, they will not be sufficient to cope with the labour which produces severe pain. No one other than the woman herself can, of course, know how much pain she is having. Only she can say whether it is constituting an experience which she finds acceptable or unacceptable. Pain, like all our sensations, is very complex and only vaguely understood. Its essential feature is, however, an experience in the mind to which we react emotionally in our own peculiar way. The significance and nature of the experience is unique to individuals.

I shall now digress to give you an outline of the mechanisms by which pain in the first stage of labour is experienced. I think you will agree that a simple working model of the process can help you and your wife understand how you, she, and we go about making the pain aspect of labour more acceptable for her. The nervous system consisting of

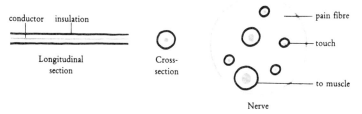

Each nerve is a bundle of nerve fibres. A fibre seen in longitudinal and cross-section

nerves, spinal cord and brain, is a system for rapid transmission of messages around the body. Each nerve is a bundle of nerve fibres which are the individual channels along which messages flash in the form of minute electric currents. Each nerve fibre is similar to the electric or telephone wires around your house. Each has a central conductor surrounded by an insulating material. The underlying functional structure of the body is organized rather like that of an earthworm—a series of horizontal slices or segments. All the messages from one slice of the body are collected together in the segmental nerve (one on either side) which passes into the portion of the spinal cord which belongs to this segment. Here messages from different types of fibres in each segment interact with each other—messages from touch, pressure, rubbing, and vibration tend to overrule or cancel out to some extent messages coming along pain fibres.

During the first stage of labour, as the cervix stretches, electric currents are generated in the nerve fibres from the cervix and these

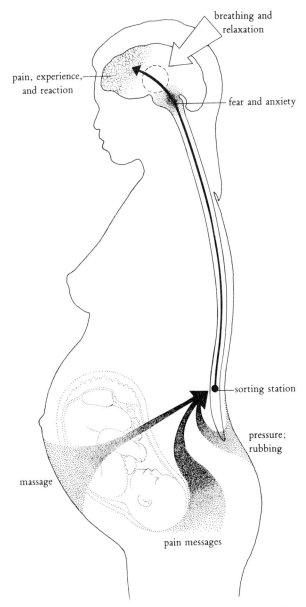

breathing and
relaxation

pain, experience,
and reaction

fear and anxiety

sorting station

pressure;
rubbing

massage

pain messages

An outline of the process by which pain sensation is developed in
the mind

transmit pain messages. These messages may be modified in the corresponding segments of the spinal cord. They then pass on along special pathways up the spinal cord to the base of the brain. Here they are further modified in a series of sub-stations in the base of the brain, before being transmitted to the interpretation stations in the surface of the brain. In the sub-stations, influences such as excessive fear and anxiety can add to the volume of the pain messages being transmitted, and reduction of fear and anxiety to reasonable levels can reduce them. The pathway from the base to the surface of the brain acts like a limited capacity pathway in that the attention of the brain has a limited capacity. There are mechanisms in the surface of the brain which interpret the messages (electric currents) which arrive in it. Pain messages are recognized as different from, say, touch partly because of the pathway along which they arrive. Your wife's 'mind' is somewhere in this surface layer of the brain. The messages have a special individual significance for her, and she reacts to them emotionally in her own way and the pain sensation she is experiencing develops a meaning or significance which is unique to her.

People's reactions to pain vary widely. At one extreme we have those who delight in pain and may have pain inflicted on themselves for their pleasure—these we call masochists. At the other end of the scale are those who are intolerant of any discomfort—these we call hypersensitive. Most people are somewhere in between and have varying degrees of tolerance. Where a woman fits into this scale of reactions depends partly on her inherited characteristics and partly on the influences and experiences which have shaped her reactions during the whole of her life to date. For instance, those women who have had very painful periods (severe dysmenorrhoea) and have managed to cope with them over the years are used to pain and therefore tend to find the pain of labour less of a problem. Those who have had pain-free lives are more of an unknown quantity, but may not cope so well with the same amount of pain.

I hasten to point out that these are only ideas which seem to me to help to explain the different ways I have observed people reacting. It is not possible to measure pain, and therefore it is not possible to know whether people who are reacting differently have or have not a similar basic amount of pain. One thing that is quite clear to me is that one cannot hope to change the intrinsic reaction pattern of a woman during the course of her pregnancy. All we can hope for at this time is to modify these reaction patterns which have been built up over a lifetime. It also seems that the power of any influence to affect reaction is greatest early in life. When your wife was in early childhood, the emotional stability and maturity of her mother, and particularly her mother's reactions to your wife's early experiences of pain, are probably powerful influences in moulding your wife's own reactions to painful stimuli. So we must accept your wife as she is and as she reacts in the difficult situation of labour. We then do our best to help her as seems appropriate at the time. There is no point in fantasizing—in imagining that any woman is going to be ideal in her capacity to cope with labour regardless of the amount of pain stimulation. Nothing we can do would bring about such a metamorphasis in anyone!

How can your wife be helped?

I will first show you how different parts of the pathways through which pain messages pass to your wife's mind can be modified by psycho-emotional and physical means. I will then describe the drug effects which may be utilized when necessary to further render the experience of the first stage tolerable. The term 'drug', by the way, means, in the medical sense, any substance given because of its beneficial effects.

There is nothing useful that your wife or anyone can do to reduce the initial pain stimulus arising from the cervix. (This stimulus is of course more intense if there is need to speed up the labour by rupturing the membranes or running an oxytocin drip.)

At the level of the spinal cord, other overriding stimuli fed in during a contraction can, to some extent, cancel out and prevent the upward

 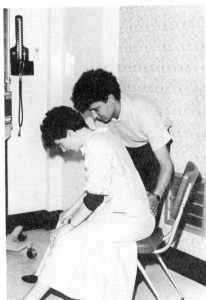

Back-rubbing or back pressure during a contraction

transmission of the messages coming from the cervix. This is why rubbing or applying firm pressure over the lower back, rubbing the abdomen, applying heat or cold, vibration, or electric stimuli may make the pain seem less.

Excessive fear and anxiety can increase the strength of pain messages passed up to the surface of the brain (at which point they are subjectively experienced). A principal aim of management at all stages is to reduce fear and anxiety to a reasonable level. This is done by gradually building the woman's confidence and trust in herself and in those who will guide her through labour, and by building her confidence in the various managements which may become necessary during the course of labour. You also need to be similarly confident as otherwise you are likely to have a negative influence on your wife's confidence. Reduction of fear and anxiety is largely a function of the antenatal classes, but it is also a function of all members of staff who are in contact with the woman during pregnancy, labour, and afterwards. All who deal with you and your wife would normally take every reasonable opportunity to explain present procedures and future events, to give advice and reassurance, and to answer questions and alter misconceptions.

The pathway from the base to the surface of the brain acts as though it has only a limited capacity. If the messages are not able to gain the attention of the brain, they cannot get through to it completely. If during a contraction your wife concentrates on controlling her breathing, tries to relax, keeps her eyes open and fixes on a spot, holds your hand, etc, the pain messages are partially blocked and she is therefore more able to cope with the contractions.

Finally, it is helpful if your wife has an accepting attitude to her labour and its pain—accepts that labour is for almost all women having their first babies a difficult, painful experience, in which she functions more on an emotional than an intellectual level. Labour has an inevitability about it which she cannot control—she has to go along with it and do her best to cope, knowing that she will get whatever additional help she wants or needs from analgesics, epidural, etc. The problem of pain at any one time depnds partly on how bad the pain is at that time, how much worse it may become, and how much longer it is going to last. We are usually not in a position to make these predictions and the midwife or doctor has to be careful not to be drawn into making them. Your wife has to try to accept this uncertainty in the confident knowledge that she will be rescued from the pain or a labour problem if and when necessary. It is very important for her to deal with labour one contraction at a time, knowing that each contraction coped with is one step nearer to having the baby. Concentrating on coping with a contraction likely to last one to one-and-a-half minutes, even though it is painful and strong, is much easier than thinking about having to cope with perhaps many more hours of labour. It is also useful in keeping labour in perspective to remember that though it is a very important experience it is still only about one day in a lifetime.

It is against the background of these coping strategies for labour that we employ various pain-relieving drugs. Analgesic drugs are most effective in the patient who has been well prepared and is well supported in labour so that fear, anxiety, and panic are minimized. The effects of analgesic drugs and good preparation work together rather than being mutually exclusive. There are large numbers of such drugs available and some institutions use an alternative to pethidine or to gas and oxygen. It is of course essential that an individual or group looking after labouring women has a restricted number of substances with which they are familiar and with which they have had considerable experience. Different individuals in a group caring for a woman should not each administer his own favourite drug, as a real muddle would result.

I am going to describe those which we use in our hospital. They are the most widely used throughout the world, though in many places epidurals or caudal blocks may not be available. They are all compatible and can all be used together without adverse effect. Many other institutions use a tranquillizing drug along with pethidine to increase the pain-relieving effect and to increase the sedative effect. We do not like this addition, as we have found that it can make a woman too sleepy and can lead to mental confusion. We prefer to increase pain relief by repeating the dose of pethidine if this is appropriate.

We use pethidine (known as demerol in the USA) as the basic, moderately long-acting substance which is the cornerstone of our pain-relieving programme. Nitrous oxide and oxygen (gas and oxygen) has a rapidly acting, short-lived effect which is suitable for limited periods. Lumbar epidural nerve blocks are an alternative to pethidine and are also available as a back-up when pethidine is not sufficient for pain relief.

I shall talk about the way we use these methods with a woman having a *first* baby. As we cannot know how much pain and individual is having, we have to rely to a large extent on her request for pain relief, but there is inevitably a need for discussion and advice. I will take up the subject of disadvantages and side-effects later.

We find that most women initially wish to try to go through labour with minimal analgesia and without an epidural block. They wish to keep the epidural block for back-up purposes. There are others who opt for an epidural block right from the start—we call this elective epidural (they elect to have it). Provided the epidural works well, as it almost always does, these women are very happy with having made this choice. They say they would not have had it any other way.

A few women having a first baby are lucky enough to have relatively easier, shorter and less painful labour and can therefore cope without analgesia or by using only gas and oxygen for one to two hours at the end of first stage. Those who decide beforehand to try to cope without analgesia can be greatly helped by good psychological support from their midwife, doctor, husband, and other attendants but the principal influence on their endeavour will be the type of labour and the amount of pain they have. They may need to compromise as the situation dictates.

Pethidine

This pain reliever was introduced in 1939, is used throughout the world, and has proved to be the best available. Its success has been that it gives those having severe pain a second chance to go on and deliver normally. It is used in the form of a colourless solution which is usually given by intramuscular injection. As muscles have a good blood supply, the pethidine readily enters the bloodstream and is carried in it to the pain pathways within the brain and spinal cord. Its effect is noticeable within 15 minutes, is fully effective in 45 minutes, and lasts for two to four hours. For the woman of average weight our dose is 125 milligrams. There are some in whom it does not produce effective pain relief within 45 minutes. In these women the dose is repeated without further delay, and is then almost always effective. We have experimented with giving smaller doses but have not found the smaller-dose method as effective. If a very experienced person is able to give individual attention to a woman in labour then the repeated smaller-dose technique can be very effective. If a woman needs pain relief, then it is essential for her confidence that the analgesic works. Her confidence can then be so boosted that she sometimes doesn't need a further dose even when the first is likely to have lost its effectiveness with time. On the other hand, not getting pain relief because of too small a dose can be very demoralizing.

We suggest that a woman asks for pethidine if or when she needs it. Her first and obvious question is, 'How will I know when I need it?' We advise that when she is having difficulty relaxing *between* contractions because she is worried about the pain of the next contraction, she has reached the time to have an injection unless she is nearing the end of the first stage. In general pethidine works better if it is given a little early rather than too late since in addition to relieving pain, it tends in some women to accentuate the emotional state of mind. Those who wait too long before having pain relief may need a bigger total dose than if they had had the pethidine sooner. This is, of course, a paradox for those who had hoped to cope without any pethidine.

If your wife gets to a stage when she is uncertain whether or not she can carry on, you should encourage her. How much you should push her to continue coping without pain relief depends on how long labour is likely to continue. The midwife or doctor may be able to give you some guidance by doing an internal examination to see how far the labour has progressed. Remember that the side-effects of one or two doses of pethidine are minimal. Most women report to us that the pethidine helped tremendously by taking the edge off the pain and that it helped them to relax between contractions. Some unfortunately are given too little too late.

The perceived effect of pethidine on pain varies from one person to another. One will say that the pain is less; another that the pain is just the same, but that it is no longer distressing her—the emotional reaction or meaning of the pain has been modified; another may say that she seems disassociated from the pain—as though it is going on in someone else and that she is looking on in a disinterested way. All are quite acceptable results.

If a woman needs more than two or three injections of pethidine to give satisfactory relief then we urge her to have an epidural.

In Britain a midwife is authorized to give the first two doses of pethidine to a woman in labour without asking a doctor to prescribe it, provided she has the necessary standing order from the doctor responsible for care.

Pethidine can be given directly into a vein, but only by a doctor. This is either as a small dose to get it acting quickly, or it can be given in the form of a continuous intravenous drip. This latter is usually a fall-back method for the woman in whom an epidural is not effective, and who hasn't had adequate relief from earlier intramuscular pethidine injections.

Gas and oxygen

For a small number of women having a first baby (and a higher proportion of women having subsequent babies), gas and oxygen is the only analgesia required for labour. This can give very effective pain relief when taken efficiently and when pain relief is required during contractions for not longer than one to two hours. It is rarely suitable for longer periods as the woman then tends to become confused, unco-operative, and out of control and takes the gas inefficiently. Her air passages also become very dry and uncomfortable. Your wife may have hoped to be one of the few women who can cope for a first baby with gas and oxygen only. Whether she starts on gas and oxygen when she feels she needs analgesia for pain, instead of having an injection of pethidine or an epidural depends on the progress of labour. If the midwife finds by internal vaginal examination that the cervix is well dilated and that the end of the first stage is likely to be within one or two hours then the gas and oxygen could prove sufficient. Otherwise pethidine or epidural will be required to give sufficient pain relief.

Gas and oxygen is often used for the last part of the first stage when more analgesia is needed and the pethidine effect is wearing off or if pethidine hasn't been required. It is also sometimes used in the second stage and at delivery.

The gas and oxygen is usually obtained through a face mask from liquefied gas in cylinders. The gas and oxygen may be each in 50 per cent concentration in the same cylinder or they may come from separate cylinders. For best results the gas needs to be inhaled vigorously at the start of a contraction. The mask is then laid aside when enough gas has been taken for one contraction. The number of breaths needed for the individual is soon determined by trial at the time. The contractions usually vary in strength so that it is better to err on the side of having extra each time. By having a break the gas effect wears off between contractions, and the woman is fully alert and ready to take the gas efficiently during the next contraction.

There is real difficulty in getting an effective concentration of gas into the lungs and therefore into the bloodstream and brain. Breathing has to be in and out against the pressure of a valve. It is necessary to hold the mask firmly over the nose and mouth to form an airtight seal so that air isn't sucked into the system to dilute it. As we are not used to breathing against a valve, extra effort has to be made and it is particularly important to make the extra effort in breathing out. Breathing has to be through the mouth and as rapidly and deeply as possible to build up gas concentration quickly as the contraction rises in strength. Breathing through the nose alone is too slow.

The first effect of the gas is to relieve pain. If it builds to higher concentration your wife may become confused and noisy. If this happens we simply get her to take a little less. If the mask is held on by another person, the vigorous breathing could result in her becoming lightly anaesthetized. It is therefore essential that she holds the mask herself. If she is getting too much, her hand falls away and there is no chance of overdose. Don't see this as your big opportunity to keep her quiet for a while by playing anaesthetist and holding the mask on! Your wife lies with the mask in her hand and as soon as she thinks a contraction is coming she holds the mask firmly and breathes quickly and deeply through her mouth for the number of breaths she has found

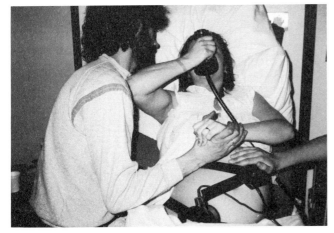

Towards the end of the first stage for this mother the pethidine is wearing off and gas and oxygen is taking over. Note the firm grip on her husband's hand—make sure your wife's nails are cut before labour, or you'll be breathing too!

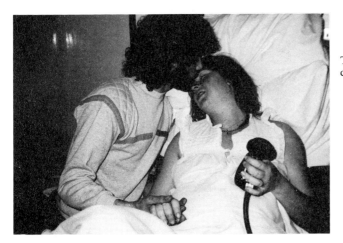

The final pain of the contraction wanes

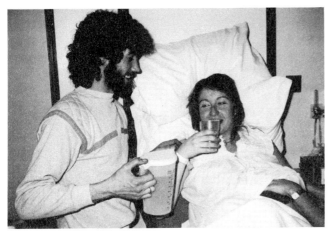

A welcome sip of water between contractions

she needs. She then lays the mask aside and concentrates on her breathing rhythm (usually in threes).

For the second stage, the gas is used in the same way if your wife thinks she needs it. As a contraction comes she takes perhaps six breaths and then pushes down with the contraction. For delivery she can breathe it continuously. As I mentioned elsewhere, many women prefer not to use the gas in second stage, provided they are sufficiently confident in the labour process and in the staff guiding them to let themselves go and to push down with abandon.

Epidural analgesia (lumbar epidural analgesia)

This technique is based on the same principle as local anaesthetic for tooth extraction or for putting stitches in a cut. Nerve fibres are like the electric or telephone wires in your house. Each has a central conducting 'wire' along which messages flash in the form of electric currents. This central core is covered with insulating material which prevents the current from leaking into surrounding tissues. A local anaesthetic agent works by temporarily interfering with the insulating property of the outer coating so that the electric current leaks out and the message is not transmitted—hence no pain message is received centrally from the affected area. If local anaesthetic is put into a wound for stitching, it stops the fine local nerves from working. In epidural block the local anaesthetic action is on the segmental nerves just before they enter the cerebrospinal fluid on their way to the spinal cord.

The spinal cord is much shorter than the vertebral column (the vertebrae joined together by the intervertebral discs to make the back bone). It and the segmental nerves entering it from below lie within protective rings of bone. Inside the bony column the spinal cord is further protected by being suspended in a bath of fluid—the cerebrospinal fluid. This fluid is contained within a cylindrical bag, the walls of which are a membrane called the dura (meaning tough or strong). Epidural means 'upon the dura'—the local anaesthetic agent is put into the vertebral canal but outside the cerebrospinal fluid so that it catches segmental nerves before they enter. It is also put in the lower back well below the level at which the spinal cord starts.

The woman lies curled up on her side or sits leaning forwards, to open the spaces between the vertebrae. A fine needle is then used to put local anaesthetic into the skin and along the line to the vertebral canal. Finally, a larger needle is pushed in gently to act as a guide for the long fine plastic tube which is then passed through the needle. The needle is then withdrawn and the fine tube taped to the skin and brought around to the front of the chest. A bacterial filter is attached and through it local anaesthetic for the nerve block is injected. The woman may feel a faint pricking as the initial local anaesthetic is put in and then a pushing sensation as the larger needle is placed. The procedure is not usually unpleasant and many women say that having the epidural put in did not worry them at all.

The idea of the epidural is to use the local anaesthetic in a concentra-

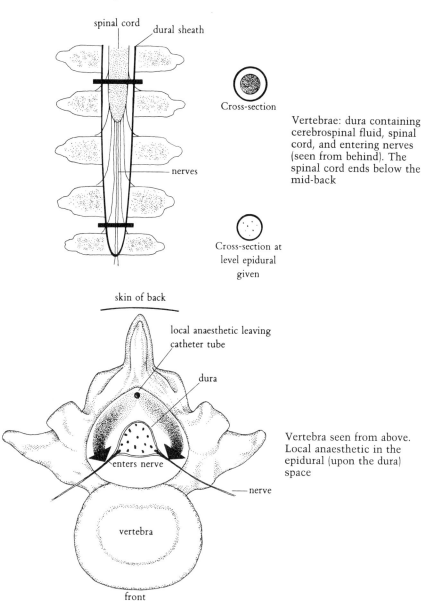

spinal cord

dural sheath

Cross-section

Vertebrae: dura containing cerebrospinal fluid, spinal cord, and entering nerves (seen from behind). The spinal cord ends below the mid-back

nerves

Cross-section at level epidural given

skin of back

local anaesthetic leaving catheter tube

dura

Vertebra seen from above. Local anaesthetic in the epidural (upon the dura) space

enters nerve

nerve

vertebra

front

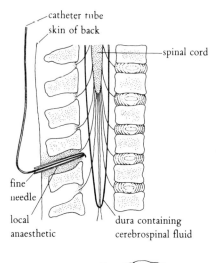

catheter tube
skin of back
spinal cord

fine needle
local anaesthetic
dura containing cerebrospinal fluid

The catheter is threaded through the needle

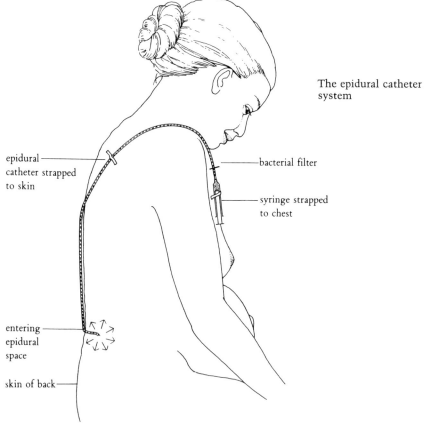

epidural catheter strapped to skin

bacterial filter

syringe strapped to chest

entering epidural space

skin of back

The epidural catheter system

tion and amount sufficient to block all the fine nerves carrying pain messages from the cervix as it is stretched during the first stage of labour. The larger nerve fibres carrying the sensations of touch, pressure, and pulling are less affected, and the large fibres carrying messages from the brain to your wife's muscles are least affected. However, it is inevitable that touch is to some extent affected—there is a varying amount of numbness. The messages to the muscles are also affected, leaving the legs feeling heavy. Bearing-down in the second stage is also less effective.

An intravenous drip is always running before the epidural is inserted, as sometimes the control of blood vessel size can be affected, leading to a fall in blood pressure which necessitates more intravenous fluid and a head-down posture for a time. After a test dose to check any effects on the individual the pain-relieving dose is given. It is fully effective within 20 minutes and lasts two to three hours. Further top-up doses are given whenever pain returns. (It is interesting to note that once a woman has had an epidural working effectively she is much less tolerant of the pain when it returns between top-up doses.)

Women who decide to have an epidural for a first labour regardless of how much or how little pain may develop do so because they quite reasonably have decided against any possibility of being martyrs to pain (they have elected for epidural). Others decide that if they need pain

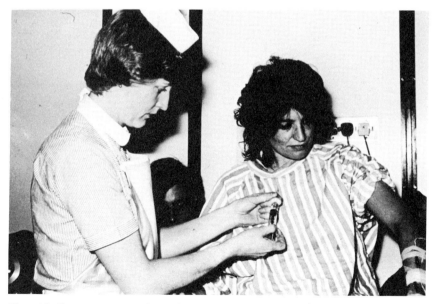

The midwife giving a top-up dose. The drip can be seen running into the arm in which the blood pressure is about to be checked

relief during the course of labour then it will be by epidural. A multigravid woman may elect for an epidural because she had one with the first labour, or because she wishes in retrospect that she had had one.

If a woman has elected to have an epidural then it is put in as soon as it is clear that labour has definitely started, or if she is to have labour induced by breaking of waters, it can be put in before the induction if this is thought desirable.

Alternatively, the woman can opt to have an epidural during the course of labour when or if she wants pain relief or can have it if relief produced by pethidine proves insufficient. A further option is that the catheter can be put in early, in readiness should it be needed later on. In practice this means it will almost certainly be used. Having a catheter in place and knowing how easy it will be to start the epidural have a powerful influence in a woman's subsequent decision to have an epidural block as contractions become painful and she is asked whether she would like it. This option thus reduces the possibility of coping without an epidural.

Once an epidural is working the woman lies comfortably on her side and can chat and read. She cannot move about the room as her legs would not support her, her blood circulation would be affected, and in addition the fetus must be monitored and the drip must be kept running.

In most hospitals women try to cope with labour with, if necessary, pethidine, often together with gas and oxygen at the end of the first stage and/or in the second stage. They have an epidural if the pethidine is not giving adequate pain relief. The slight conflict in this option is that it is less easy to get an epidural correctly placed and working well when a woman is having a lot of pain in advanced labour and therefore less able to lie or sit as still during the insertion. An epidural is more likely to work well if given early, but this removes the option of trying to cope without it. It is always a good idea before starting the epidural for a check to be made by vaginal examination that the first stage isn't nearly complete. Pain is usually most severe as second stage is approached. In this case it can be too late for an epidural for first stage, and the woman may well decide it isn't necessary when she knows the most painful part of labour is nearly over. Analgesic use must always be in association with obstetric assessment of progress.

The ideal is that the epidural works well for the first stage and then stops working for the second stage to allow the woman to push more effectively. This can be arranged to some extent, but the first-stage epidural invariably continues to work to some extent in second stage, thus reducing the efficiency of the woman's second-stage effort. It is also difficult to predict even after vaginal examination how much longer the first stage is likely to last and it is often unreasonable to suggest not having a top-up of the epidural in the hope that the end of first stage is near.

Some women prefer not to feel anything in the second stage and at delivery. For them an extra top-up is given in the sitting position at the start of the second stage to allow the solution to run down and affect the lower nerves coming from the pelvic floor—the lowest part of the birth canal and around the outlet of the vagina.

It is interesting to note that some women who have had a longish labour, for which they had an effective epidural, comment that they found labour tedious and boring. They say they were, however, excited when the baby came. Once again, it is getting there rather than going there which is the fulfilment of labour.

Sometimes the attendant staff will advise a woman to have an epidural for her particular labour. This may be because she has raised blood pressure or is having an immature baby (premature labour) and sometimes when she is to deliver twins or a breech. If it is clear quite early on that labour is very painful, and that further doses of pethidine will be unlikely to bring sufficient relief, then again epidural is likely to be advised. When epidurals are not available then relatively large doses of pethidine, perhaps including an intravenous drip of pethidine, can meet the need. The disadvantages of the large doses can be accepted and viewed as a reasonable compromise in these difficult circumstances.

By no means all hospitals have a reliable epidural service or one which can be called on 24 hours a day, seven days per week. A specially trained anaesthetist almost always starts the epidural and remains responsible for it. It is not uncommon for a midwife trained in the technique to give the top-up doses. Sometimes an obstetrician with suitable training and experience substitutes for the anaesthetist.

If a woman with an epidural working in labour needs an emergency caesarean section (*see* also Chapter 20), then the epidural can be augmented to give satisfactory anaesthesia for the operation. Whether or not this is done depends on whether the woman wants this form of anaesthesia for the caesarean, how well the epidural is working, how urgent the need for caesarean is, and how experienced the available anaesthetist is at augmenting the epidural so that it is suitable for caesarean section.

You may wonder why everyone doesn't opt for an epidural for pain relief in labour. There are several answers: many women don't need it and cope very well with labour perhaps with one or two doses of pethidine. Some don't like the idea of the whole procedure and its disadvantages (see pp. 115–16); they feel that they are not doing it all themselves and can feel they are being taken over by tubes and technical devices. They would also prefer to feel more of the sensations of the second stage and of the baby being born. It is therefore much better for them to see how they get on in labour and, reserve epidural for back-up purposes.

For the woman having a lot of pain, the big advantage of epidural over the larger number of pethidine injections which would then be necessary is that she is awake and alert when the baby comes. Those opting for or needing an epidural for a first labour, should remember that second labours are usually much less distressing. They can still have the experience of labour without epidural the next time if they wish to.

Caudal analgesia (caudal epidural analgesia)

This is essentially the same as lumbar epidural analgesia in its mode of action and effects, except that it is very unlikely to be suitable for a caesarean section. The main difference is that the needle is inserted just above the tailbone and passes upwards, rather than being put into the small of the back as with the lumbar epidural.

What are the disadvantages of pain-relieving substances?

First of all there are side-effects. These are effects of the drug other than those for which the drug is given. One has to compromise and accept side-effects to obtain the benefit of pain relief. Neither pethidine, nor gas and oxygen, nor epidurals has any proven effect in hastening or slowing the progress of the first stage of labour.

Pethidine

Some women may *feel sick* or vomit—if this becomes a problem another substance can be given to counter this side-effect. Vomiting is, by the way, not uncommon in normal labour at the end of the first stage, even when drugs have not been given—it empties the stomach ready for the pushing of the second stage.

Some can become quite *sleepy* for a time after an injection—this wears off and isn't necessarily a disadvantage. This calming or sedative effect can reduce anxiety, and your wife rests, relaxes, and perhaps sleeps between contractions. If a woman needs pain relief, then her ability to co-operate and do her breathing and relaxation is much better after pethidine than it would have been without the pethidine. If your wife is sleeping then you watch the monitor for the start of a contraction or notice your wife stirring with the contraction. Say firmly and clearly to her: 'breathe, and open your eyes and focus on your spot'. This stimulates her attention. Otherwise the contraction might near its peak and catch her unawares. Obviously the ideal situation would be for her to have good pain relief and yet remain fully alert, but a compromise has to be accepted. A point that isn't always appreciated is that even in women not needing analgesia the state of alertness diminishes as labour progresses. They too can feel tired and often wonder how they will have the energy to push in the second stage. They are functioning on a more basic level. (Some doctors and hospitals fairly routinely give a tranquillizer such as sparine with the pethidine, and some of the supposed sleep-producing effects of pethidine noted by women in labour are due to the tranquillizer. It has been our experience that it is better to give higher doses of pethidine together with personal reassurance than added tranquillizer. We are very much against the use of tranquillizers in labour. Their use is however, a matter of opinion.)

Pethidine is a very effective pain reliever, and in my experience most women having it think it is a great help, but it doesn't eliminate pain. If a woman feels in retrospect that it hasn't helped her adequately, then she perhaps has had an insufficient dose, or needed an epidural. Your wife may say that the pethidine hadn't helped when you, at the time, could clearly see that it had. With a strongly advancing labour, the pain would have been much worse without it. Some are rather light-headed for a time but should just drift along with this feeling. Occasionally women go into second stage more quickly than anticipated, so that they are more sleepy in second stage and at delivery than they would have liked. This is a nuisance but acceptable if needed for pain relief. Often the lack of recall is more due to associated tranquillizer.

Some women develop a rather romanticized expectation of the magic moment of birth, and if they have had an injection of pethidine can feel they have missed out on something and blame the pethidine. In fact most women, regardless of whether or not they have had a pain reliever say that delivery brings a great sense of relief and don't comment much about the magic. Some do find the moment of birth an overwhelming, unbelievable experience but this is not necessarily influenced by whether or not they have had pethidine.

WHAT ABOUT THE BABY?

Almost all substances given to a pregnant or labouring woman enter the baby's circulation, and pethidine is no exception. It depresses the baby's initial respiration, and when the mother has had *several doses* of pethidine it can occasionally have an important influence on the initiation of the baby's breathing after birth. However, much the most common reason for a slow start to breathing, or even not starting at all without expert assistance, is that the baby's vital functions have become depressed because the baby has become rather short of oxygen. This is a possibility in any labour and pethidine can add to this effect. This can be vitally important if delivery takes place away from expert

Equipment of this type used for resuscitating the baby is often in the delivery room or nearby

assistance, but is of little account in a good hospital unit. Such a unit should always have someone on hand who can resuscitate the baby. This means sucking mucus from the baby's mouth and throat and then giving oxygen via a face mask, but preferably through a small tube passed into the baby's windpipe. The pethidine effect can be reversed to some extent by injecting an antidote into the baby's circulation through the vein in the cord or indirectly through an injection into the muscle. If no more than two doses of pethidine have been given within eight hours of delivery, any pethidine effect on the initiation of respiration is insignificant.

When a baby does not suckle well after delivery it is easy to blame any pethidine which has been given within a few hours of delivery. Pethidine can impair the baby's early suckling, but one tends to forget that many babies whose mothers haven't had any drugs seem to take a day or two to get the hang of suckling. While everyone will be pleased if the baby suckles well at the start, there is no good evidence of any long-term significance as far as the establishment of breast-feeding is concerned. In my opinion the effect of pethidine is, like the possible effect on initiation of breathing, of little importance in the long run. The main disappointment arises when a woman attaches too much significance to the baby's early hours of life. Remember that 'nothing's either good or bad but thinking makes it so'.

Babies whose mothers have had pethidine may be a little more sleepy in the first few hours, or for longer if larger doses have been needed for pain relief. I think that these effects are an acceptable compromise in order to make labour reasonable for the mother. If, on considering the baby's immediate interests, one tries too hard *not* to give pethidine, there can be major disadvantages for the baby. Women who have been severely traumatized by suffering unrelieved pain in labour can 'subconsciously' blame the baby for their suffering. They can be so disturbed by the experience that they fail to form a warm emotional relationship with the baby. I have come across the problem in the past and mention it to temper the enthusiasm of those who wish to withhold analgesia. The most important thing from the baby's point of view is that he has a loving mother.

GAS AND OXYGEN

The action of nitrous oxide is very short-lived. Apart from the mother's occasional slight feeling of nausea, or temporary confusion if too much is taken, there are no important problems. Any effect on the baby is minimal. Other gases such as trilene which are sometimes used instead can have a more cumulative sedatory effect on the mother and a little sedatory effect on the baby. As mentioned earlier, gas and oxygen is not a suitable anagesic for prolonged use especially for a first labour.

EPIDURALS

When an epidural is working well it gives complete relief of pain and is therefore ideal from that point of view. Almost anyone is a suitable subject for an epidural but there are a few exceptions, such as those who have had a back operation and those who have a tendency to bleed, or are currently having a problem with blood loss. Epidurals are undoubtedly a major advance in the provision of pain relief, as anyone who has been having severe pain and has then experienced the complete relief of it with epidural will testify.

The effectiveness or success-rate or epidurals varies with the skill and experience of the anaesthetist. Sometimes the epidural may be only partly successful but brings some relief. Or it may 'take' on one side only, which is not very helpful. Some 5–10 per cent are unsuccessful even with a good anaesthetist. It is therefore necessary that a woman opting for an elective epidural should still have learned how to cope with labour without an epidural just in case she needs to. Occasionally, the epidural works quite well but the effect isn't maintained with later top-ups. The epidural can then be re-sited and will then probably start to work well again. Sometimes it doesn't, and it may then be necessary to rely partly on pethidine or a similar drug.

The epidural imposes a marked limitation on mobility—the woman has to lie on her side or perhaps adopt a semi-sitting posture for a time, but with relief of pain she may no longer feel the need for mobility. I have reservations about women adopting the semi-reclining position after an epidural has been set up. It does not speed labour more than lying horizontal on either side but it may result in less satisfactory blood flow to the placenta.

When second stage is established, women usually have an overwhelming desire to bear down with the contractions. This reflex bearing-down may be weak or absent with an epidural, and the woman may not feel any contractions. It is not necessary to bear down until the head is low in the pelvis near the outlet. By this time, the bearing-down reflex may have returned but if not, an attendant can feel contractions with a hand on the abdomen or note them on the monitor, and tell your wife when to push. In general the second stage tends to be less efficient in women who have had an epidural in that the baby delivers less readily and it is therefore more often necessary to assist delivery with forceps or the ventouse. This is usually a fairly straightforward procedure for which the epidural usually gives good anaesthesia after a further top-up dose given in the sitting position.

Once an epidural is working it is essential that the woman does not lie flat on her back, as this accentuates any tendency to lower the blood pressure.

Occasionally when the guide needle for the small tube (catheter) is being inserted it can prick the dura (the membrane that encloses the cerebrospinal fluid). This leads to a slow leak of cerebrospinal fluid and to an alteration of pressure relationships and a slight drag on ligaments, so that about 1 per cent of women having epidurals can be troubled by headache for a few days. This can be a very nasty headache, only partly relieved by lying flat and by simple pain-killers. If the possibility of a leak is recognized, the headache may not start if the woman is kept lying flat for a day or so after delivery.

During labour many women with an epidural find they are unable to pass urine and need to have a catheter inserted to empty the bladder. This can occasionally continue to be a problem for a few days after delivery. Epidural makes this problem more likely.

Disadvantages for the baby are the increased likelihood of a usually easy forceps delivery; possible fetal distress because of reduced oxygen supply if the blood pressure falls for some time; and sometimes increased floppiness of a few hours because of absorbed local anaesthetic. The advantages of more efficient pain relief offset disadvantages.

13 *Monitoring during labour*

The purpose of observation is first to check on the progress of labour, secondly to check on the condition of the baby, and thirdly to check on the condition of the mother. I shall deal with these separately, although they are going on concurrently and are to a large extent interdependent. The intention is to allow the natural processes to follow their course. They are, however, observed closely with a view to assisting when it appears that they are not performing satisfactorily in the best interests of mother or baby.

The progress of labour

The progress of labour is judged by observing the general reactions and demeanour of your wife; by feeling the descent and rotation of the baby's head through the lower abdomen; by intermittent vaginal examinations to check the descent of the baby's head and the rate of dilatation of the cervix; and, in the second stage, by observing the rate of descent of the baby's head down the birth canal and through the vulva. The apparent strength and frequency of uterine contractions are taken into account in assessing the significance of the rates of change in these observations.

Sometimes the nature of the action of the uterus cannot be followed sufficiently well by simple observation and feeling the uterus through the abdomen. The strength, frequency, and co-ordination of the uterine contractions can then be recorded by a small pressure-measuring device, placed within the uterus after the membranes have ruptured. Before this, the frequency of uterine action and to some extent the co-ordination and strength of contractions can be recorded by a pressure gauge held against the abdomen by an elastic belt.

Progress is related to average standards of expected rates of progress allowing for a fairly wide range of individual variation. For instance, we would expect your wife in her first labour to be progressing in the first stage at least at a rate likely to have the labour completed within 24 hours of labour becoming established. We would not expect the second stage to take longer than one to two hours, and we would expect significant progress over a period of 20 minutes during second stage.

Sometimes progress in early second stage, before the head comes down to the vulva, can be less certain and a longer period of observation is reasonable provided the fetal welfare is carefully monitored.

The condition of the baby

One can never afford to be complacent about the condition of the fetus during labour. Most times the baby is reliably supplied with oxygen and essential food substances via the placenta, but, as I have described in the section on the placenta, the fetus can be walking a tightrope. Evidence that the baby may be getting into difficulties is obtained by observing the colour of the liquor, by observing the fetal heart rate and rhythm, and sometimes by studying the acidity of the baby's blood obtained by a scalp sample.

If the baby's bowel content, which is green because of the presence of bile, appears in the liquor, the baby may have had an epidode of oxygen lack, and more careful observation is called for. One of the reasons for breaking the waters, once labour is established and the cervix is dilating, is to detect this meconium staining of the liquor should it occur.

The baby's heart rate and rhythm may be checked most simply by periodically using an ear trumpet or a hand-held electronic device. It is particularly important to use these during and just after contractions. This method may overlook some of the more subtle but important changes, and may be late in detecting the more obvious fetal heart-rate changes.

Various other types of recording apparatus are available. These may produce a continuous recording of the fetal heartbeats, and time them in relation to the urterine contractions. For instance, in deciding the significance of an episode of slowing of the fetal heart rate, it is important to determine the time relationship of the slowing to uterine contractions. The recording may be by means of a fetal heart rate detector on the mother's abdomen or, more accurately and reliably, by attaching a small metal clip to the baby's scalp after the membranes have ruptured. Both can have long leads so that your wife can move about a little and perhaps sit in a chair. Alternatively, signals can sometimes be sent to the recorder by a transmitter without the need for leads. (This is called telemetry.)

The monitor may give a continuous display of the baby's heartbeats. You may be worried when the baby's heart-rate is slowing, but the midwives or doctors will explain what is going on. You may find the monitor quite useful in that it will often show the start of the contraction before your wife feels it. You can warn her to start the breathing and to be ready for the contraction. Most men and their wives find monitor-

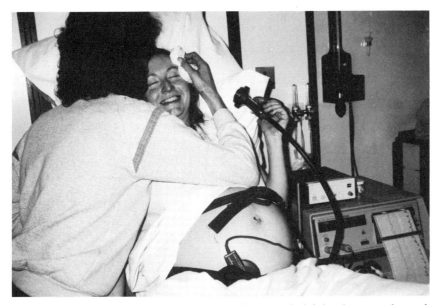

The fetal heart rate is monitored by the lower appliance to the left-hand trace on the graph paper. The upper appliance is a pressure detector giving the right-hand trace showing contractions of the uterus. Nearing the end of first stage after pethidine and gas and oxygen (between contractions)

ing a positive and very reassuring feature of their labour management.

A sample of blood can be taken from the baby's scalp and analysed for its acid content. This is usually to be more sure whether or not the fetus is short of oxygen when this has been suggested by alterations in the heart-rate pattern. This blood sampling can be done only when the cervix is sufficiently open and the baby's head has come some way down the birth canal, so that the tubular instrument used can be passed without much difficulty or discomfort for the mother.

You will hear the terms 'fetal distress' or 'suspected fetal distress' mentioned. They are used to indicate that there is evidence that the fetus is or may be short of oxygen. This is what we are looking for in fetal monitoring. The term distress would suggest some emotional upset, but this is not implied—we don't know about this aspect of fetal functioning and are considering the possibility of physical harm arising in the form of brain damage due to oxygen lack, if it is in fact present and allowed to continue.

The mother

The aim is to keep a check on your wife's welfare and also to see that she has good care, reassurance, encouragement, and pain relief as necessary.

We check her pulse and blood pressure at regular intervals, as these may reflect a developing internal problem, such as bleeding. We observe the volume of urine produced and check it. If your wife is short of fluid it may be necessary to run fluid continuously into a vein (called an intravenous drip). There is at present some disagreement about the need to use glucose fluids in this way.

As a general principle, nourishment should not be given by mouth during labour. Oral intake is restricted to sips of water or sucking water from a sponge to stop the mouth from becoming too dry. However very early in labour, or when labour is not well established, something readily absorbed such as white bread (or toast) and tea are often taken. During labour the stomach tends to be inactive. Anything taken by mouth tends to stay in the stomach and is not absorbed. It may well be expelled around the end of the first stage or the beginning of the second stage, when vomiting is not uncommon.

The real danger is that any woman in labour may unexpectedly develop an emergency situation for which a general anaesthetic is required. If there is food or fluid in the stomach this may come back and enter the lungs during the anaesthetic, with serious consequences and even death. Glucose drinks or sweets taken by mouth and then later inhaled are more irritant than most. In any case food or fluid taken by mouth once labour is established is unlikely to be of benefit because of poor absorption.

Reassurance and encouragement are the province of everyone in contact with your wife during labour. The foundation for this reassurance is laid during the antenatal period by all those with whom your wife is in contact. This process of confidence-building is a specific concern of antenatal classes which should, wherever possible, be centred in the obstetric hospital where the delivery will take place. Staff likely to be present in labour should take some part in the classes. The hope is that the labouring woman and her husband will trust the staff caring for them. Likewise this trust is reinforced by good physical care, explanations, and reassurance. You and your wife need to sense in the staff an attitude of genuine concern for your welfare and you in turn need to be able to trust their professional judgement. The management of the pain and distress of labour is a central and very important issue in labour management, but much more complex than a superficial examination would suggest.

14 *Your role in labour*

Being present in labour and for delivery is not for every man. Nowadays the pressure to conform in this way may be very strong. It is presented as though absence will result in a failure of the family and an inability to relate effectively on an emotional level with the baby as he grows. This is quite obviously an exaggerated view and quite clearly not true. Sometimes, for one reason or another, a man is not able to attend. Others don't wish to be there and occasionally a wife does not want her husband there. However, we have found over the years that most men who have participated have been pleased to share the experience and have found delivery amazing and miraculous. Almost always the wife has appreciated her husband's presence. If you decide not to attend, or for one reason or another cannot be there, your wife will be well cared for. Even if her mother or other suitable companion is not available, most hospitals have someone who can spend most of the time with your wife if the midwife happens to be too busy to do so. It is an important general principle that women shouldn't be left alone in labour.

The traditional role of the husband in labour has tended to be peripheral—mostly restricted to boiling water when confinements were at home (probably as a diversionary activity, as I cannot believe that vast quantities of boiled water were required). With the move to hospital confinement, men tended to be even more left out. Over the last 30 years their involvement has become gradually established and the valuable contribution you as the father can make is now widely recognized. I can picture you thinking, 'How could I? I'm squeamish and may faint at the sight of blood.' I can assure you that very, very few men faint at delivery although we prefer to have you seated and always like you to feel free to walk out if you want to. We do keep an eye on you! Being there and actively part of the process is somehow easier than reading a book or watching a film about labour.

What can you do?

Well, first of all, merely being present can be a great comfort to your wife. Hospitals can be rather frightening places and your wife will appreciate having the company of someone with whom she is at least vaguely familiar. You are very much part of the patient side of the team and won't be expected to run the labour or undertake any tricky procedures, but you will be included in explanations and will have the chance to ask questions. Remember that your wife is inevitably anxious in labour so try not to ask questions or make comments in her presence which are likely to heighten her anxiety. You can, however, help her to understand what is said and can ask questions on her behalf if necessary.

You will have the same anxieties as your wife about whether the baby will be normal, whether he is getting enough oxygen, and whether everything is going along satisfactorily with the labour. The staff will aim to be realistically reassuring and to keep you in the picture, but above all you must be patient.

Initially your wife is likely to be cheerful and pleased to be in labour. She says, 'This is all right—I can cope with these contractions and don't need to do any breathing.' It's a good idea to start her practising at this stage. When she is well on in labour she may find the contractions very painful even when she has had one or two injections of pethidine. She will say, 'I had no idea labour would be as painful as this. They don't tell you it's going to be like this.' Your attitude and demeanour are important. It's not helpful if you are too sympathetic and say: 'Oh dear. I don't know how you're coping. I'm so sorry, I had no idea it would be so difficult for you. I couldn't do it', and burst into tears. Nor will saying 'Look here, you've been to all the classes and you know all about labour. Stop making all this fuss and get on with it.' Somehow, you have to strike a middle course. Try to be optimistic without being too euphoric, in the knowledge that labour will be over one of these days. Recognize and accept the difficult task your wife has in coping with labour, don't argue with her, and be guided by the staff. There is an important difference between firm, encouraging reassurance and bullying. Say, 'I know it is difficult for you, dear, but you are doing really well—do your breathing, keep your eyes open, and focus on your spot.' We all like to have our efforts appreciated.

In the earlier part of labour you are keeping your wife company. You may like to walk about together or go to sit in the waiting area. You can encourage your wife to relax and practise her breathing. During a contraction she may find it more comfortable to lean forwards against you or a wall, or on the back of a chair. Don't talk to her during the contraction other than to encourage her with her relaxation and breathing. It doesn't matter that she grips your hand tightly, and it doesn't matter how well she manages to relax. The important point is that her mind is diverted towards concentrating on trying to relax.

When your wife is well on in labour, she is usually not very interested in conversation. She will just want to lie on her side in bed, or sit in a chair trying to relax and concentrate on coping. You may think you are doing a good job in keeping the conversation going only to be told rather abruptly to 'shut up'. In labour, your wife is very much the centre of attention—she will be very intolerant of any other conversations going on so whatever you do, don't start discussing the latest cricket score with the medical student, or last night's television play with the midwife.

Your wife may have a sudden emotional outburst in which you are told fairly frankly and in unusually plain language just what she thinks of you. She may say it's all your fault and say, 'this is the last baby I'm having'. One girl, in reporting her labour experience to a class just about to have their first babies, said she was pleased to report that she didn't say she wouldn't have another, but said she had to admit that she said to her husband towards the end of the first stage: 'Do you think it's too late to have an abortion?' Your wife may say she can't cope and can't go on any longer and may burst into tears. It is important that we find out just what the trouble is. If it is pain that is worrying her, then she needs more pain relievers. However, it may just be a sudden emotional release of tension and anxiety which have been building up in her. Sometimes her main need is for reassurance and, with this, she will settle and continue with labour. Such outbursts must, of course, be taken seriously, but not taken too much at face value.

When you have been told that the first stage is nearly over then your wife may be encouraged by being reminded that it won't be too long before she is in second stage and able to push. However, be careful, as so often this end of first stage can be prolonged and your reassurance can wear thin. It is perhaps more relevant in the second stage when progress is slow and your wife is getting fed up to remind her that it won't be too long before she will see the baby, but again don't be too optimistic.

Sharing the experience of labour, whether good or bad, usually improves mutual understanding of events. In the past, when men were largely excluded, it could be very difficult for a man to appreciate just what his wife had been through, and this could adversely affect a relationship. For the wife to be unable to communicate about a traumatic or difficult experience sometimes drove a couple apart. A man sometimes, not unreasonably from his point of view, thought his wife was making too much of a song and dance about what he had been led to believe was a pretty normal process. When he visited afterwards and found his wife sitting up in bed in a new nightdress, with a lovely baby alongside, he could be left with an entirely unreal perspective of labour. When he left her early in labour, she was fine, and when he saw her afterwards, she was apparently fine. Sometimes women bottled up their feelings or resentments as they didn't want to upset their husbands.

From the man's point of view, it can be more traumatic to pace up and down outside for hours, worrying and fearing the worst, than to be present throughout—warts and all. Don't feel you cannot leave your wife's side for a few minutes to get a cup of tea and have a short walk if there is someone else to stay with her. This can give you a break. You see that the world is continuing and you regain your perspective on life. It's a good idea to take in some food. You won't be much use to your wife if you become hungry, irritable, and exhausted from anxiety and lack of nourishment.

Whether or not you are present at the time of delivery is again something which you and your wife need to have discussed and thought out beforehand. A woman may expect not to cope very well at delivery and feel that she could lose face with her husband, or that he may no longer subsequently find her sexually attractive. A man may be concerned about seeing his wife in pain and distress or he may be concerned about the sight of blood, which is inevitably part of the delivery scene. Fortunately, in practice, these anxieties don't seem to be too important. The point of being at the delivery is not so much to watch it as to be at the head of the bed holding your wife's hand, encouraging her and mopping her brow, and being there to share the experience of your baby's arrival. This can be a real highlight in your lives. However, if you are not at the delivery all is not lost, as is sometimes suggested.

The idea of epoxy glue-type bonding with the baby at delivery is a grossly exaggerated myth (*see* Chapter 17). It is far better to follow your inclinations about being present at delivery having first found out what it may be like and after talking to your wife about the whole exercise.

After it's all over, don't be too disappointed if your friends who haven't been at a delivery are not too interested in the minute details. It can be difficult for them to tune into the spirit of the occasion. If you are not there, expect some of your friends to tell you you've missed the greatest experience of your life!

If you are one of the few unlucky ones who has not liked being there or been upset by some aspect, try to talk to your wife about it. You may find that she, in the midst of it all, has been less affected than you think. Often a woman says that she could not have managed if her husband hadn't been there, even though he may have felt relatively impotent. Your wife will want to be in touch with your real feelings and to help you cope with the occasion. Don't make out that the experience wasn't traumatic for you if it was, but you can also point out the positive features as you saw them.

You may be interested in the results of a questionnaire survey of 730 fathers reported in 1972 by Pawson and Morris from the West London Hospital. The first table shows the main ways in which they thought they had contributed, and the second table the main feature identified by the 251 fathers who found something distressing. The overwhelming majority were glad to be with their wives and singled out sustaining morale as their principal function and they saw themselves as allies.

You may be all too aware of a feeling of helplessness and frustration that you are not able to do enough to help your wife, especially in coping with her pain in first stage. Remember that everyone present is inclined to feel this way. There is very little that you can do personally apart from mopping her brow to cool her, giving her sips of water or ice to suck, giving her lip salve, holding her hand, perhaps rubbing her back, telling her when contractions are on the way down, and encouraging her with breathing. Although you may feel relatively useless all these things will be important to your wife and your very presence will help her.

	Number of fathers
Helped to sustain morale	615
Provided physical comfort	147
Helped with breathing	112
Furthered husband/wife relationship	56
Wife appreciated sharing experience	23
Helped pass instructions	11
Eased pain	10
Total	730

Of these, about one third identified upsetting features:

Wife's pain	116
Helpnessness	25
Attitude of staff and lack of attention	24
Blood and placenta	23
Episiotomy	18
Forceps	9
Emergence of head	9
Total	251

When labour is over, don't forget to praise your wife for her efforts. She will have done her best, often under very difficult circumstances, but will need reassurance of this. There is no standard against which she or you can judge the 'performance'. Each is a unique experience. Talk it all over and look for the good points and the happy ending and try to be positive about what has happened.

15 *The arrival of your baby*

Many parents feel joy, elation, and excitement at this thrilling and emotionally moving event. You and your wife may well be moved to tears. An individual birth and the reactions to it can seem miraculously unique even to attendants who have been present at many births. They often feel some of the emotion and are in tune with the spirit of the occasion.

It is not uncommon for your initial feelings for the baby to be much stronger than those of your wife. Especially if her labour has been long and exhausting, her reactions may be more of relief—'thank goodness that's over'. She may be so emotionally drained that she has nothing left for this initial reaction to the baby and she may feel quite flat.

There is, however, contrary to most people's expectations, a wide variety of reactions to the baby's arrival. Quite apart from the emotion and general excitement induced by the occasion, don't be surprised if neither you nor your wife has a feeling of deep emotional attachment for your baby. She often seems like a stranger; and you don't really know her. The emotional bonds between you and her develop at varying rates—it may be several hours, days, or even weeks before one or other or both of you feels towards her as you expect to feel. Sometimes part of the joy and excitement you display at the birth is put on for the benefit of others present, as you think they expect this reaction. You may be secretly disappointed at the appearance of your baby—especially if you think there is a strong resemblance to a relative you are not keen on! You may be disappointed that the sex is not in accordance with your strongly held expectation. Sometimes a man or woman is worried and guilty about not feeling an immediate emotional bond with the baby. You may think you are lacking in parental instinct and perhaps unsuited to the parental role. These are not unusual reactions and don't seem to be of any significance provided you don't take them too seriously. It is easy to develop an over-romanticized view of this over-emphasized bonding concept. Your feelings become stronger as you care for and get to know your baby.

Your wife is usually given the baby to hold as soon as he is seen to be

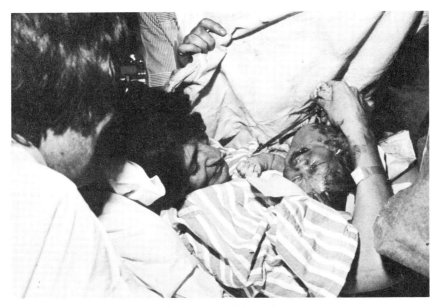

Thomas—just delivered

Thomas with his father

Thomas aged three. A joy to his grandfather and all his family

breathing satisfactorily. Some women like the baby to be delivered on to the abdomen but others find this too messy. Unless your baby starts to breathe and cry fairly quickly, it is important to clear the nose and mouth of excess mucus as this can occasionally block the airways. There is usually no urgency about tying and cutting the cord if the baby is at the same level as the placenta. If, however, he is above it, as when he is on your wife's abdomen, then the cord should be clamped fairly quickly. The pulsation of the cord is the pulse of the baby's heart pumping blood from him into the placenta. The blood returns to him passively—and in this case against gravity. It is easy to see how he can lose part of his blood volume into the placenta if it is not clamped quickly when he is on your wife's abdomen. If he is below the placenta then there is a tendency for blood to accumulate in him and for his blood to become too thick. The amount of blood left in the placenta after tying the cord should ideally be the same as it is at delivery.

By the way, the placenta normally separates from the uterine wall as

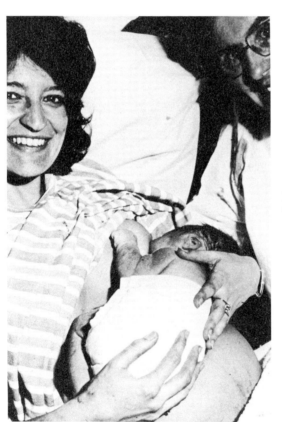

Many babies will breast-feed straight after delivery

the uterus contracts with the delivery of the baby so there is no continuing oxygen exchange with the mother.

At delivery the baby is wet and can lose heat very quickly if he is not dried fairly soon and wrapped. Ideally there should be a heat radiator moved over him as he lies in your wife's arms or on her abdomen. If, in a warm room, his skin is against the skin of his mother's abdomen and chest and he is covered after being dried, then heat loss isn't usually so much of a problem.

If breathing is slow in starting then it is a first priority for him to be resuscitated.

If your wife is going to breast-feed then she (and you) perhaps with the assistance of the midwife, may wish to put him to the breast within a short time of delivery. Many babies suckle readily at this time but others don't seem interested. Don't be too disappointed as many babies who subsequently feed very well are slow starters. You too will, of course, be given your baby to nurse and admire.

Your baby will be weighed, measured, and examined, at least superficially, and wrapped warmly.

After delivery of the placenta and any stitching that may be necessary, your wife will be washed, and given a fresh gown, and you will probably be given a cup of tea. You can then have some time together with your baby provided there are no problems. You will probably already have popped out to make 'phone calls. Some people like to bring in a bottle and have a celebratory drink, but don't press it on the doctors or midwives as they are working and may have any sort of emergency at any time.

16 Bonding: mother–baby or father–baby attachment

It is progressively more widely recognized that all reasonable steps should be taken to have the mother and baby together as much as is practicable following delivery. The same applies to the father, brothers, sisters, and other involved family. The baby should be delivered on her mother's abdomen if this is what the mother would like, and given to her to hold, fondle, and caress and examine as soon as compatible with the baby's safety. This applies also at delivery by caesarean section under epidural anaesthesia. If a general anaesthetic is used, mother and baby should be together as soon as is practicable. After a birth, the parents should have time with their baby in private for, say, 30–60 minutes. This should allow them to undo and inspect their baby and hold her naked (provided there is a heat shield) and put the baby to the breast if they wish.

On post-natal wards, mothers should have their babies with them as much as they wish, and should always have ready access to them. On the other hand, a nursery must be available as there are times when a woman wants the baby to be there. If the baby is unwell or immature, parents should be encouraged to have as much access as is reasonable. The mother can almost always hold her baby for a short while before she goes off to the special or intensive care unit. Once there, the mother and father should be encouraged to visit and touch and hold their baby as much as is reasonable. If the mother has had a caesarean delivery, but is otherwise well, she can soon go along in a chair and she can always have polaroid photos to look at. Some units are able to project pictures of the baby from the special care unit to a bedside monitor screen, so that the mother can view her baby continuously.

In the unit in which I work, grandparents and siblings have, for many years, been given free access in the neonatal unit and we have not found a problem with infection.

There has been, in recent years, a considerable increase in discussion about the importance of early and prolonged interaction between mother and baby in establishing the quality of future parenting. The principal stimulus for this discussion came from the work of the Americans, Klaus and Kennell, who from 1972 questioned the tendency

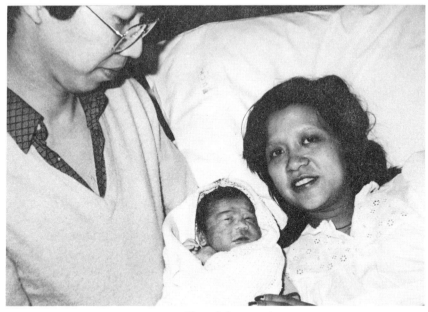

Ten minutes old, with his proud parents

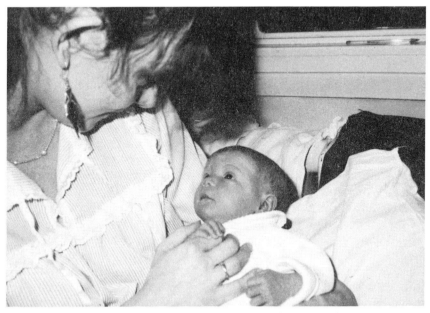

Four days after caesarian delivery, Nancy hopes for her next breast-feed

most marked in the USA to whisk the newborn baby off to a nursery. The baby was then kept away from mother and father for 6–12 hours. There was much emphasis on protecting the baby from possible infections by this artificial separation. Along with this policy, there was a tendency to release the baby from the nursery for feeding at regular intervals, and to whisk him back to 'safety' without delay! Klaus and Kennell initially compared two groups of mothers one month after birth, scoring them for the amount of attention, interest, and soothing behaviour they displayed towards their babies under controlled conditions. The first group experienced standard practice for their hospital—i.e. whisked away, etc.—while the second had extended contact with their babies. This extended contact, which was perhaps decidedly modest by the standards of most of the world's mothers, consisted of one hour of holding the baby within three hours of birth, and an extra five hours of contact every afternoon for the next three days. The extended-contact group achieved clearly higher scores for their attachment behaviour.

These initial observations pointed the way for further studies which have been reported. It has long been known that for sheep, cows, horses, dogs, and ducks, the first few hours after birth of the young are crucial for establishing normal maternal behaviour. If attachment does not occur soon after birth, the baby animal is rejected and starves. It is not hard to see how, with this information, the imaginations of some enthusiasts have run riot a little, as they have been stimulated to theorize that for humans there is a similarly critical period. Some have even suggested that, if the mother does not have the baby immediately after birth, all will be lost.

This exaggerated concept of an epoxy-glue-type bonding process occurring within the first few hours is quite clearly untrue, as your grandmothers and mothers will tell you. They and you have survived the rigours of the nursery-type system. The theory behind this nursery-oriented regime has been shown to be unfounded and, applied in this fashion, certainly had the potential for harming mother-baby relationships. This does not imply that there is any harm in a baby being in a nursery for a time, if her mother is sleepless and tired and would like her there. Klaus and Kennell were surprised and taken aback that their observations should have led to extreme views, with the implication that any necessary separation was very likely to be harmful.

The subsequent work did confirm noticeably beneficial effects on mother–baby relationships of this extended exposure for *first–time* mothers of low socio-economic groups, but did not demonstrate any heightening of the bond in women of the middle and upper socio-economic and educated groups. The theory is that maternal conditioning of the educated, middle, and higher economic groups had already implanted attitudes making early maternal–baby touching of less importance in later development of mother–child relationships. From experiments performed, there is no evidence that the middle-income mother, deprived of seeing her baby for the first 12–24 hours of its life, will enter into a damaged mother–child relationship.

Mother–child, as well as father–child, bonding or attachment is a continuing process lasting many years. No amount of early contact can substitute for the devoted care and teaching of the first 20 or so years.

There is no doubt that the establishment of the love relationship—of warm emotional bonds between parents and child, is of crucial importance for the healthy development of the family. Multiple influences are brought to bear on this developing relationship; some influences such as the mother's care by her own mother, family and cultural influences, are fixed and relate to the past. Others relate to the present—behaviour of doctors, nurses, and relatives; separation of mother and child, and many others. The bonds are not fragile things which don't establish unless conditions are optimal, and there is no reason to worry when all influences are not positive. Great anxiety about having optimal influences is likely to be very counterproductive. However, there is no need for scientific proof before applying commonsense measures.

The commonsense guide is to have everything done as naturally and peacefully as is compatible with the safety of mother and baby. Unless there is some good reason for doing otherwise, the mother should have her baby to cuddle as soon after delivery as she wishes, and then during her hospital stay to have the baby with her as much as she wants. However, women should not feel unable to let the baby out of their sight for a short while when they are very tired and in need of sleep. If you wish to be at the delivery and to handle your baby soon after birth, this cannot be other than good. However, if you are unable to, or don't wish to be at the birth, then there is no good evidence that you should be pressurized or feel in any way inadequate for not having been there.

17 Developmental defects (congenital abnormalities) and damage around the time of birth

Most biological errors giving rise to a malformation of the fetus, or responsible for miscarriage in the first four months of pregnancy, occur in the very earliest stages of development of the embryo. Less often the embryo starts off developing normally, but may be damaged a little later. An example is when the mother develops rubella (German measles) during the first three months of pregnancy. Fortunately, such later problems are rare. The fetus which starts developing normally is fairly safe up to the time of birth. Malformations (early developmental defects) are discussed in Chapter 7.

Congenital abnormalities

These are several defects of varying importance with which your baby may be born. They may be temporary and trivial, as with cephalhaematoma (see below), treatable medically, or permanent and

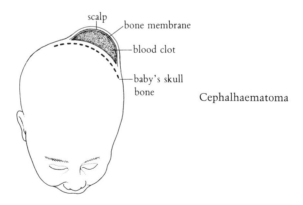

Cephalhaematoma

untreatable. Some, such as the strawberry-type birthmark, disappear with time. Others, like the port-wine stain, are permanent and may or may not be treatable. Cleft lip (often with cleft palate) can look horrifying yet be very effectively amenable to surgical correction.

Cephalhaematoma is bleeding between a bone of the baby's skull and the skin over the outside of the bone (the periosteum) produced by the rubbing movement which occurs during labour and delivery. It causes a prominent lump about 5 cm in diameter towards the back of the top of the head and to one side. This gradually absorbs, leaving a rim of extra bone on the outside which in turn gradually absorbs over a few months. The most serious congenital abnormalities are those which are not amenable to successful correction, are compatible with survival, and yet not compatible with the child leading a reasonable life. Fortunately these latter are rare but are the ones that all parents fear.

If you are one of the 1 or 2 per cent of couples whose baby is born with a significant abnormality then you should be told about it and its implications as soon as the problem is known. Not surprisingly you will both be shocked and upset and will need sensitive help in accepting your bad luck. It is so much better if you are together at this time and can see your baby together. Questions will spring to your minds but there will not be ready answers to some of them. Doctors and nurses will be on hand to help you. (Please refer also to the section on stillbirth and neonatal death (p. 193).)

Usually when a fetus is found on testing in late pregnancy (usually by ultrasound and/or X-ray) to have a serious abnormality not compatible with more than a brief neonatal life, then it is appropriate to induce labour as in a case of fetal death. The situation will of course be fully dicussed with you both.

Damage around the time of birth

Minimizing or elminating damage around the time of birth is the prime concern of obstetric and neonatal care. Above all else the wish of parents is to have a healthy baby who has the potential to develop to his capacity mentally, emotionally, and physically. Doctors, midwives, and nurses aim to provide all reasonable care and assistance to help achieve this end, while at the same time being fully aware of the needs of you, the future parents. Sometimes your wishes and desires have to be modified to achieve these ends—for example, accepting and being thankful for a forceps delivery when you have set your hearts on having a normal delivery.

The principal underlying problems which contribute to a baby being damaged around the time of delivery are immaturity of the fetus (being born early) and inadequate oxygen supply to the baby's brain during labour, immediately after delivery or in the first few days of life. Immaturity predisposes the newborn to failure of oxygen supply. Sometimes we can postpone threatened early delivery by using drugs which reduce the activity of the uterus. We can also recommend general measures, such as extra rest, which may reduce the likelihood of the early onset of labour when there is a predisposition, as, for example, in twin pregnancy. Monitoring techniques are used to detect early evidence of oxygen starvation in pregnancy and in labour, and caesarean section or forceps can then deliver the baby more quickly when this danger threatens.

If left to nature, most babies would deliver normally without damage. There would, however, be some potentially healthy babies who would be badly damaged, or even die, because of shortage of oxygen. There would also occasionally be others who would become infected in the course of prolonged labour. A few would completely fail to deliver and die if the bones of the pelvic passage were too close together to permit delivery. To minimize these problems, birth is often assisted when a situation which may be dangerous for the baby develops. The assistance itself may very occasionally be responsible for damage. In the relatively rare situation when the shoulders get stuck after the baby's head emerges, the fairly complicated manoeuvre sometimes necessary to deliver the baby can occasionally result in damage to the nerves to the arm, or a bone in the arm can be fractured. This is, of course, preferable to the baby not delivering at all and therefore dying, and is acceptable, though very much regretted, if the result is that most babies thus trapped are delivered healthy.

Many people still worry that a forceps delivery is likely to damage the baby. Prior to 1935 when the sulphonamide drugs became available, there was no specific treatment for infections. We were mostly dependent on the natural processes for overcoming infection. This meant that caesarean delivery was very dangerous and to be avoided if at all possible. At that time, 10–15 per cent of women who had a caesarean delivery for obstructed labour which had been in progress for more than 24 hours died of infection. For the sake of the mother, every effort was therefore made to avoid caesarean section by delivering the baby vaginally with forceps. The serious and often permanent damage sometimes suffered by the baby was regarded as acceptable when there was no reasonable alternative for delivery.

Nowadays, caesarean section has become a very safe operation, and we don't hesitate to employ it to avoid the type of forceps delivery which used to be a necessity and which resulted in the dubious reputation of the forceps. Forceps are now used only when they can safely deliver the baby. Difficult forceps deliveries have therefore been replaced by caesarean delivery. Any anxiety about forceps delivery is nowadays almost competely unfounded.

Caesarean delivery has now become so safe for several reasons, which include the wonderfully effective antibiotics available to combat infection; the safety, effectiveness, and availability of blood transfusions which have improved enormously in the last 40 years; the great improvements in anaesthetic agents and the techniques and skills of anaesthetists; and the improvement in surgical techniques and materials.

If you are unlucky enough to have a baby born abnormal either because of a developmental defect or because of damage sustained about the time of birth, you will both then need to talk about the tragedy with both your obstetrician and your paediatrician. You will both feel sad, angry, and disappointed and somehow guilty that you haven't produced a perfect baby. If your baby survives then you will need extra help and understanding and will have many problems to face. You may try to put on a brave front, but don't be ashamed to be upset and sad. You will need to talk over together all your feelings and reactions and to comfort each other. Read through the sections on stillbirth and neonatal death. Yours is a loss—a loss of the perfection you had hoped for.

18 The post-delivery interview

In the few days after delivery we try to let each woman have a chance to talk about her labour and delivery. She is encouraged to discuss her feelings and recollections. Any misunderstandings can be resolved to enable her more easily to become reconciled to her experiences. The likely influence of her current pregnancy and delivery on future pregnancies, or the likelihood of repetition of problems, is discussed. Problems likely to be encountered in the weeks after delivery have already been introduced during the pregnancy and can now be discussed further. There should be no question of assessing your wife's performance, because for one thing her experience and her reactions to it are quite unique and cannot be meaningfully compared with those of someone else.

In most hospitals you yourself are not usually given this opportunity to go over your experience of the labour with one of the staff. When we have done this we have found it has been very helpful in allowing you to come to terms with your experience and perceptions of the labour. At least you can talk them over with your wife. If you were there, you may have been more in a position than she was to take in explanations given during and immediately after the labour. Often you will at that time have had the opportunity to ask questions and to talk with doctors and midwives. Any discussions, which you or your wife have subsequently need to take place with someone who has genuine knowledge of the details of your wife's labour experience. We hope that in the future these very useful discussions will become usual not only when serious problems have arisen but also when labour and deliveries have been less spectacular and basically have gone very well.

Some people your wife meets in hospital or afterwards ask such questions as: 'Did you do it naturally?', 'Did you have an epidural?', 'How many doses of pethidine did you have?', 'Were you monitored?', 'Did they induce you?'—as though these are important matters!—your wife will have had what was reasonable for her under the circumstances. There is no point in making the whole process into a competition—it sometimes seems as though her character is being assessed. You may be asked similar questions and feel you have to justify her having had pain relief. Unfortunately childbirth is increasingly more often regarded as a competitive activity, and this inappropriate element cannot be entirely eliminated. Fortunately most people are not like this and will be delighted the baby has arrived and that he and your wife are well.

Do not be surprised if your wife seems to want to go over her labour and delivery experience frequently. She will often want you to clear up her uncertainties. Her concept of the passage of time in labour will probably be highly inaccurate. Above all she will want to be reassured that she has done well.

19 *The three months after delivery*

In the first three months after the birth you have a time of major adjustment which brings much pleasure, happiness, and fulfilment. On the other hand, like so many of life's experiences, it is by no means always plain sailing. There will be times of deep disappointment and even despair. There may be times when your relationship as a couple will come under extreme pressure. There is no simple answer to this danger. Sometimes you have developed an unreal picture of this so-called idyllic time. Sometimes you have difficult social or housing circumstances. Some people have a happy contented baby, whereas others can be unlucky and have one who is colicky and crying. At times, usually in the early weeks, he will not settle and can keep on crying in spite of feeding, winding, changing, cuddling, etc!

At one day old Roxy gives little hint of future disturbed nights

The early weeks

Most find the early weeks difficult, demanding, frustrating, and tiring, but at times relieved by feelings of joy, happiness, and fulfilment as they see their baby gradually start responding to them. You may be surprised to find that as early as two weeks after your baby's birth you will both be feeling it is difficult to imagine life without him. Each new development brings great pleasure and often excitement. It is the good feelings which endure. Hence the over-optimistic descriptions given by those well through this early period! You are most likely to get a realistic picture from a couple with a *first* baby just three months older than yours. Many women keep in touch, at least on the telephone, with one or two of the women they have met in hospital. Remember that parenting is competitive. Just listen to what marvellous managers your parents were when looking after you!

Fortunately, the difficulties of the early period almost always resolve with time. It is useful to think of pregnancy as lasting 12 months rather than the usual nine. If you can be reasonably settled as a family by three months after the birth, then you are well on course. Keep saying to each

Now we are three

other that under the circumstances you are doing well and that every thing will get better—and it will. Give your wife plenty of encouragement and praise and make sure you support her in her way of looking after your baby, even if you are not 100 per cent sure you agree. There is usually no right way. Having confidence in her own ways is of paramount importance for your wife's success as a mother.

I shall talk about some aspects of the first three months. Many of these are most evident in the first few weeks after delivery. Although it is easier to discuss them in more or less separate sections, they are in real life all going on together and very much interrelated.

Your wife's moods may at times be quite baffling and you can wonder why her coping abilities seem to have waned. Many influences which vary in relative importance between individuals, and from time to time in the same person, are operating. The following are those which are usually most important. There are changes in the body's hormones; there are often after-effects of pregnancy, labour, and delivery; tiredness from lack of sleep; loneliness for the woman used to being with people at work; frustration at the apparent endlessness of this 24-hour-per-day job; there is also happiness and fulfilment and the admiration and love of friends and relatives; but anxiety about the baby's welfare; about feeding; about parenting ability; perhaps about taking up a career again and the arrangements to be made to allow it; and by no means least, there is you and your needs, demands, reactions, moods, and coping abilities. There is also the ongoing and changing relationship with you and with friends and outsiders.

First of all, hormones: these are chemical substances which come from the body's glands and, in pregnancy, from the placenta. Whereas the nervous system is the fast method of communication between you and the different parts of your body, to keep them co-ordinated, the hormones form the slow but longer-lasting system of control and co-ordination. These hormones flow around in the bloodstream and affect every part of the body including the brain, in which moods are controlled. The effects of the hormones of pregnancy are seen in miniature in the phases of the menstrual cycle. The principal hormones are called oestrogen and progesterone. You cannot have failed to notice their effects. In the five to ten days before the menstrual period many women become much more difficult to live with than usual. At this time, your wife may seem irritable and quarrelsome for no obvious reason. Following the menstrual flow all is often sweetness and light, and your relationship prospers. If you chart the family rows in relation to the phases of the menstrual cycle, you may be surprised to find that they are concentrated premenstrually. I would, however, hasten to add that males have been known to be unreasonable and that when you are beastly to your wife, it is isn't quite fair to blame the consequences on her hormones!

In pregnancy the placenta acts as a huge hormone factory, with the object of bending all your wife's body's functions to the interests of the pregnancy. It takes the body some time to adjust to these new demands. As the hormone levels rise in early pregnancy we have variable combinations of nausea, vomiting, metallic taste in the mouth, strange appetites, excessive tiredness, irritability, etc. As pregnancy progresses, the mood tends to settle but most women remain emotionally labile—that is, your wife is more sensitive and may get down in the dumps easily and be unexpectedly tearful, or be really elated and happy when circumstances haven't changed much. Some women are more calm and peaceful when they are pregnant. Nesting instinct, vagueness, and forgetfulness are all partly hormone related.

After delivery, the placenta is lost and there tends to be an overswing to under-production of these hormones. Hormone levels produced by the ovaries may not return to pre-pregnancy levels for several months. This relative hormone deficiency can underlie the emotional instability of the early weeks. You may first notice this as the 'third day blues' which strike between three and seven days after the birth. Your wife becomes uncontrollably depressed and tearful for a few hours or a day or two. You may leave your wife one evening in quite good spirits; you arrive perhaps a few minutes late the next night with only two bunches of flowers, to be greeted with tears and recriminations. 'Why can't you be like other husbands? They manage to arrive on time!' If you aren't prepared for this type of reception you can wonder what on earth has gone wrong. If you react as you might reasonably do in other circumstances you can easily precipitate a major family crisis! Remember, however, that your wife's feelings are real to her and not to be trifled with. If you were to pat her on the head and say, 'There, there, I know it's only your hormones, dear', you are likely to be lynched. You have to take the situation seriously, yet not too seriously. Your wife may go on to tell you how inconsiderate the staff are, and how she can't stand the woman in the next bed, etc. A tricky situation, but at least you have been warned!

I don't want to overstate the role of 'hormone balance' as quite frankly we don't know how important it is in causation of 'third day blues', even though it sounds a plausible theory. Other influences operate. However, the relief afforded by this emotional blow-out *does* seem to be important —we regard it as an almost essential and inevitable feature of the puerperium (the few weeks of recovery and adjustment to the birth of the new baby). There tends to be a build-up of anxiety towards the end of pregnancy, followed by the stresses and strains of labour and delivery, the excitement of the new baby, the sleepless nights and the anxiety about feeding and about the baby's welfare. All probably contribute to a further build-up of tension which then floods out.

In our hospital we have for many years had a much-appreciated custom. We suggest that the couple has a meal out together the night before going home. This may be the last opportunity for some time and is a good chance to become re-acquainted! If your wife is breast-feeding she can arrange to leave some expressed milk, and the nurses will cope. Perhaps you could enquire whether you could have this excursion from your hospital.

Experience would seem to suggest that the ideal situation for your wife when she comes home with your new baby is that she is relieved of all responsibility other than caring for the baby. She really needs to be cosseted and able to relax to establish feeding habits and her relationship with the baby. This concept was possible in the days when she used to go home to the bosom of the extended family. It is unfortunate in some ways that in our more fragmented society this is usually no longer the pattern. There is often just you and your wife. I am not suggesting that there aren't many advantages in having your independent set-up, but this is a time when assistance is needed. It is important if possible to round up a peaceful relative or friend, or to employ a 'mother's help', and to arrange beforehand for her to come and take over the care of the household. Your wife's mother is sometimes ideal, but I fear your own mother is less often the right person because of the inevitable rivalries. Often the supporting domestic role falls mainly on your shoulders. Unless you are already well into it, you should try to make the metamorphosis and pitch into domesticity with a will. It is ideal if you can arrange to have one or two weeks off work to coincide with your wife's return home. It can be almost impossible to arrange a fixed time well in advance, as the time of delivery cannot be accurately predicted, but your employer may accept the uncertain timing of your absence.

You will often be surprised how your previously well-organized wife gets into a real muddle and can't seem to schedule her life. When you come home in the evening you are likely to find her sitting in the middle of the floor crying with the baby. The dishes are still in the sink and nothing seems to have been done. You may tentatively ask what she has been doing all day to be told, 'Looking after your b—— baby, and it's your turn now!' This is not the signal to carry him up and down the room a couple of times and then decide that it is really his mother that he wants, or that he needs feeding when your wife has just fed him!

Babies have frequent mood changes. Many kinds of distress can be soothed by the age old ploy of movement—especially rocking. Think about investing in a rocking chair. Your foot suffices for manipulating a rocking cradle. An automatic rocker is becoming available.

Domesticated husbands are not new. Even though not the father, Joseph was required to cook breakfast. (From the early fifteenth-century alterpiece in the Pfarrkirche Bad Wildungen W. Germany)

The artist appreciated that there were times when even Mary rejected her baby. (From Master Bertram's Grabower Altar-piece 1379–83)

Conflict of advice can be a major burden. This is, I am afraid, inevitable. Everyone is an expert on the subject of baby care and baby management. You and your wife will get plenty of apparently well-meant advice from friends and strangers whether or not you ask for it. In most hospitals there is some attempt to provide consistent advice by having a co-ordinated general approach to some aspects. Despite this, the personal views of nurses, doctors, cleaners, domestic staff, and visitors tend to obtrude. Your wife will also tend to have her own ideas based mostly on the way she herself was brought up, but perhaps modified by reading or some other learning process. She then tends to ask advice from everyone until she finds someone who agrees with her. This is why, rather than give advice, it is better for an attendant to find out what a particular woman wants to do and, provided it is part way reasonable, to agree and reinforce these ideas and, at most attempt only slight modification. In this way your wife's confidence increases and it is this self-confidence which is so important. It is self-evident that when there are so many firmly held views and theories about baby care, there is no correct way. This is why the individual woman's feelings about what is right should be reinforced and supported.

This shower of advice continues when your wife is at home as people tell of their own experiences of babies—of course their baby was always a joy and so easy; rarely cried; smiled; gained weight at the right speed and was no trouble. Your poor wife contrasts this with the reality of your baby who is wakeful, cries so much of the time, won't feed easily, and demands her constant attention.

It is worth remembering that mothering and parenting generally are competitive and that most people remember only the pleasant things from the past, especially when their cometence is at stake. You can be the rock on which your wife can anchor. Encourage and praise her efforts and try not to be too demanding yourself. Discuss details which are not going well and, perhaps most important, have your own authoritative book to which you can both refer when in doubt. Spock (*Baby and child care*, published by Star Books) is still one of the most popular and practical. Read together the relevant passages which you have already been over during your wife's pregnancy, and try to ignore outside advice, except for that of one or two wise counsellors.

Tiredness from lack of sleep can be a real problem for your wife and also for you. One of your important jobs can be to protect her from too many visitors. It is likely that your wife will be looking forward to holding court in hospital as all and sundry come to pay their respects to her and the new baby. But they tend to stay too long, and your wife soon becomes tired and is too busy to be entertaining visitors. It is sometimes better tactfully to discourage most visitors other than the immediate family and close friends. You will also both want to have time together in the hospital, without the company of visitors. Just say that your wife has been losing sleep and is rather tired but that she would appreciate a visit later on perhaps when she is at home. Your wife will not have slept on the night before delivery and is usually too excited to sleep the first night after the birth. She will then find she has very broken and fairly sleepless nights as her baby indulges in frequent feeds and there are other disturbances in the ward. She will be lucky to average more than four to five hours sleep at night during the few weeks ahead. It is also important, after she comes home, to suggest to visitors after a short stay that your wife is tired and needs to get to bed early. Otherwise some people are liable to sit around the whole evening without realizing the problem. Encourage your wife to have a daytime nap when your baby is asleep. Your wife should also try to get some sleep after the early-evening feed before the baby wakes for the next feed. Especially in the next few weeks she is likely to be up again in the small hours and then again at 6 or 7 o'clock in the morning. As each feed and change can take at least one hour initially, there isn't much time left for sleep.

You also can become tired if you have an unsettled baby—you may be required to drive him around in the car most of the night so that he doesn't disturb the neighbours. The motion of a car seems to be ideal for getting a baby off to sleep, but he may wake each time you stop at traffic lights! Your baby will also find the motion of a pram, or being carried in a baby sling or carrier, soothing so you may be quite busy. If your baby is bottle-fed, you will soon become an expert at feeding, winding, etc!

If your wife has had a caesarean birth or some difficult problem during pregnancy or labour or at delivery, it will take her longer to recover her energy. A caesarean section is a major abdominal operation to which the body doesn't always take kindly. Most people having a similarly extensive abdominal operation such as a hysterectomy expect to be looked after and to convalesce for a couple of months. A woman after caesarean section is expected to care for and feed her baby. She usually goes home 8–10 days after the delivery, but is in no state to take over the household. If your wife has a caesarean delivery you must make sure she has assistance—in some areas you can, if necessary, arrange through your Local Authority for a home help to do housework, shop, etc.

Despite the many difficulties, there will be times when you enjoy the early weeks (at least in retrospect). There is no better way to get to know your baby and for the family to become a threesome than for you to become really interested in playing with and looking after her. Most hospitals will be keen to train you in napkin changing and bathing, and your interest can be stimulated in other ways. Babies are much more aware and reactive from an early age than they have in the past been given credit for.

If you hold your baby about 25 cm from you, looking directly at you, there is now good evidence that she can see your face from very soon after birth. It is perhaps a rather indistinct circle with a couple of blobs for eyes, but it seems to have for her a recognizable shape. She will hear well and will respond to your voice. You will soon learn how to soothe her. Babies who are frequently talked and sung to, cuddled and fondled, seem to thrive on this extra attention. Babies can smile from birth and within a few weeks smile in response to you. You will be endlessly fascinated as the weeks and months slip by and she gradually develops her abilities. It is well worth reading about the various stages with the aid of a book on child development, and perhaps logging the changes you see. Remember that rates of development are very individual and there is a wide range of normal. Within this range landmarks haven't much value for predicting future intelligence and achievement, so don't get too carried away or your friends will label you a baby bore! With a first baby each step in development is like magic because you haven't seen it

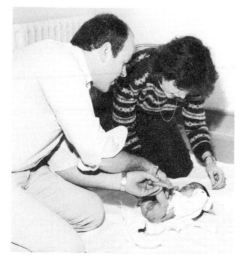

Nappy changing may not be every father's cup of tea

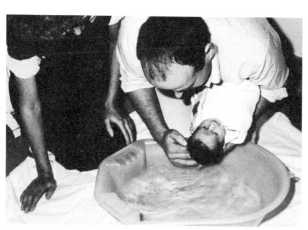

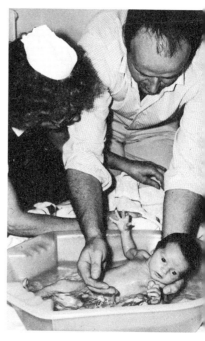

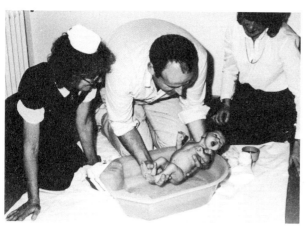

Induction into the mysterious art of bathing. This couple has continued to prefer the floor, as there is nowhere for baby Rhys to fall!

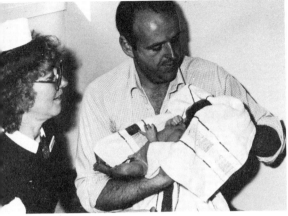

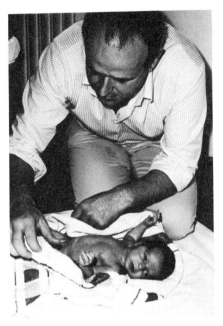

Next comes drying, which includes all the nooks and crannies

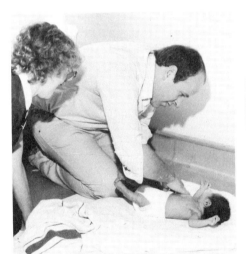

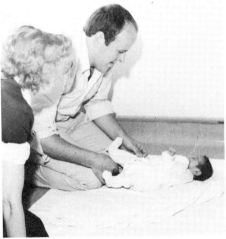

. . . and then the complications of dressing

before—with second ones these steps are still fascinating but can be taken more for granted.

Breast or bottle?

The decision as to whether your baby will be breast- or bottle-fed is one in which you should take more than a passing interest. Your wife's chances of becoming a successful breast-feeder can be very much affected by your attitudes and encouragement. If she wishes to breast-feed and is successful with it then there are considerable advantages. If, on the other hand, a strong distaste for breast-feeding persists after the birth then bottle-feeding can, with care, be fully adequate. Incidentally, with bottle-feeding you can be more involved in the direct process of feeding.

If your wife is going to breast-feed successfully, then she and you must be fully committed to it and feel there is no other way. Some have no difficulty in establishing feeding, whereas many who go on to successful feeding have real troubles at the start. There can be cracked and sore nipples which may not heal easily; the breasts can be swollen and painful when milk production first starts (engorgement), and occasionally a breast infection develops. Many women are anxious about whether or not the baby is getting enough whereas the amount taken from a bottle is easily seen. You don't however know how much an individual baby needs so this advantage is more apparent than real. The best guides are the baby's contentment, the character of bowel motions, and weight gain. Your midwife, health visitor, or doctor will help and guide you with any problems. Your help and support of your wife's endeavours are so important. Such innocent remarks as, 'Is it really worth it?', 'Wouldn't it be much simpler to stop trying and upsetting yourself and just put her on the bottle?' undermine confidence and resolve.

The paradox is, however, that while really not entertaining the possibility of failure, you must both be prepared to accept that at times breast-feeding just doesn't work—some women have nipple characteristics which make feeding almost impossible, while a few women just don't produce enough milk. There does come a time when your remarks as quoted above become most appropriate. The end-point for giving up the intention to breast-feed is always arbitrary and can only be arrived at by discussion between yourselves and with your advisers. Unfortunately, we have seen occasional cases where the commitment to breast-feeding is such that a couple won't give up even when the baby is obviously undernourished. There can even be continuing weight loss rather than the intermittent or steady gain one looks for. Some of these

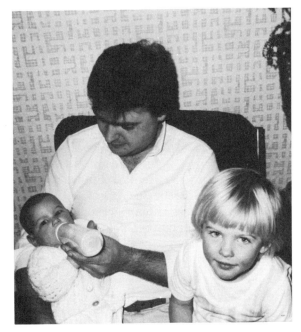

Dad is already an expert
(and Lisa wanted to be
included)

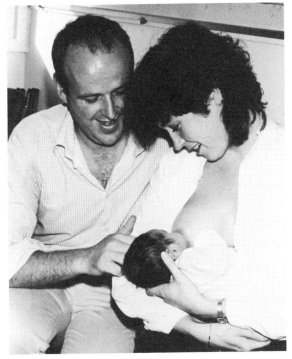

Successful breast-feeding

undernourished babies don't cry much, hence the value of occasional weighing. Quite clearly the primary aim of any feeding is to feed the baby. We must sometimes be prepared to accept and adjust to defeat, even when it can mean temporary loss of face or self-regard.

The principal advantages of breast-feeding are:

1   The milk is of a consistency most suited to the baby's nutritional needs—cows' milk can be cleverly modified by modern manufacture and mixing, but it is never quite the same. This may not matter too much normally but is increasingly important if the baby becomes ill with a raised temperature. He can then have difficulty eliminating the extra salt in cows' milk. Your wife's milk is initially a yellowish colour (colostrum), but when lactation is well established it is watery bluish white in colour. It looks watery because the principal protein of human milk is albumin which is, like the white of raw egg, colourless, whereas the principal protein of cows' milk is the white casein. Breast-milk has, however, better nutritional value for your baby than the white cows' milk.

2   Breast-fed babies are less likely to develop infections—especially gastroenteritis. The antibacterial antibodies in the colostrum are protective. It and the milk encourage a better balance of the numbers of the normal bacteria in the intestine. Cows' milk can harbour harmful bacteria if it is contaminated after sterilization and then not kept adequately refrigerated. Bottles and utensils used in preparation can become infected.

3   Breast-fed babies are less liable to some allergies and there is even a suggestion that they may be less likely to develop coronary artery disease in later life. Breast-fed babies are certainly less liable to be overweight though they may not win baby shows when scales are a principal means of judging!

4   Breast-fed babies have less digestive upset and colic and never develop constipation even though they may pass large soft stools quite infrequently.

5   Close physical contact with the mother is inevitable. This contact seems to be important in the growth of the love relationship between mother and baby and in the development of the baby's feeling sensitivites. (On the other hand, if a woman is breast-feeding out of a sense of duty but is really hating it, then perhaps she would be better bottle-feeding. When bottle-feeding is the order of the day, then this close contact is maintained by cuddling during feeding and at other times. I don't think the bottle-fed baby is necessarily disadvantaged in this respect.)

6   Breast-milk is always available and at the right temperature when mother is about. It is so much easier for night feeds and is certainly more convenient when travelling. It is always possible to accumulate enough expressed milk over the day and keep it in the fridge so that your baby can have breast milk from a bottle if it is not convenient for you and your wife to take the baby if you are going out for an evening. When the baby is a little older he can be given a bottle of cows' milk on these occasions.

7   Some people worry that breast-feeding will alter breast shape, but any alteration is really because of pregnancy rather than the feeding. Sometimes those who have been concerned about small breast size are altered for the good. (Small breast size is, by the way, practically never a bar to breast-feeding, as the gland tissue grows during pregnancy and the amount of milk subsequently produced is largely regulated by the baby's demands.) For those with moderate or large breasts, sagging can usually be prevented by wearing a good supporting brazzière during waking hours in pregnancy and during breast feeding. For many, this means also wearing a lighter bra for sleeping.

8   Breast-feeding may reduce the likelihood of the woman developing breast cancer in later life.

9   Women breast-feeding often use up extra stored fat, so that their figures can return sooner.

10   Finally breast-feeding can stimulate erotic feelings in some women with an earlier resumption and enjoyment of sexual intercourse. On the other hand some women can feel guilty about such feelings and may even give up breast-feeding because of them.

Many people have heard about four-hourly feeding. Your wife may imagine that she will feed the baby and then put her down and have most of the four hours to get on with the other things she has to do. This can be the pattern after two or three months (or earlier if she is lucky), but most babies require frequent feeding—sometimes up to 12 feeds per day between one and three weeks of age when the milk supply is becoming established—so it's not hard to see how one feed will run into another. Demand feeding (which means feeding your baby more or less whenever she cries for feeds) usually works out best in the long run. Milk production is increased by more frequent feeding to match baby's needs. Giving extra milk from a bottle is sometimes advised but removes the stimulus for extra milk production. Too much effort to adjust the baby on to a regular feeding schedule can end in a shambles with a fraught, frustrated baby and an unhappy, tearful mother. Patience, confidence, and compromise are needed. Sometimes a dummy (pacifier) can be useful to comfort your baby for a short time while your wife finishes a task or can be used as an aid in settling a fretful baby.

Cot death

You and your wife will continue for some time to be anxious about your baby. This is normal, particularly with first babies. You will no doubt have heard of cot deaths—cases where an apparently healthy baby is found dead usually with no satisfactory subsequent explanation. Fortunately they are relatively rare, occurring in only about 1 in 1000 babies. It seems there is very little you can do to prevent this tragedy occurring. If your baby is unwell there may not be much evidence of it. He will sometimes fail to feed and seem listless. If in any doubt consult your doctor. Try not to overheat or underheat your baby (the former is more likely). If in doubt, slip your hand under her clothes; the baby's skin over her chest should be pleasantly warm. If it feels sticky she is overclothed, and if the skin is cool she is underclothed. Remember, however, that babies are tougher than they seem—they are nine months old at birth and are to a considerable extent parent-proof.

In the early weeks you and your wife will spend quite some time creeping in and out to see that your baby is still breathing. Many times people have reported poking the baby to make sure she is still alive and then spending the next two hours regretting having done it, while trying to get her back to sleep again!

Your relationship with your wife

You are changing to add the role of father to that of lover, and you will have to share your wife with your baby. You may have come to regard yourself as of some importance in the family until your baby arrived! You can then begin to wonder just where you fit into the heirarchy. For want of a better term, we call it a jealousy reaction. I am not suggesting that you are likely to plan an attack on the baby, but you may find yourself reacting excessively to small irritations. For example, you should try to resist the temptation to suggest changing the baby from breast to bottle when the main problem is that you are feeling rather neglected. Without being fully aware of it, you may be reacting to the notion that the baby is taking up so much of your wife's time and attention, and he is even commandeering your territory by breast-feeding. You may find yourself marching out and slamming the door when on reflection you are not too sure what has sparked the outburst.

On the other hand, there may be some substance to your reaction in that it is very easy for your wife to become so wrapped up in her pregnancy and the baby that she can forget she has a husband. Your wife can speak about 'my baby' and seem to forget that she is 'our baby' and that you love her, too. She can unconsciously relegate you to the role of fetcher, carrier, and provider so that occasionally you can begin to resent her as well as the baby. Sometimes a woman will have already in late pregnancy made the mistake of turning too readily to her mother and other relatives to discuss matters such as the decoration and furnishing of the baby's room. She may be too anxious to please her relatives and parents in choosing a name. The obvious answer is that you both make a special point of discussing as honestly as you can with each other your feelings, fears, anxieties, and plans for your baby. In this way, you maintain a sensitivity to each other's needs. You must both adapt, accommodate, and make allowances for each other. We all need time to adjust to changing circumstances—but adjust we must. You will no doubt accept that the relationship between your wife and your baby must always to some extent be exclusive, but your perhaps lesser role in caring for your baby in an involved physical way comes much more easily if you have been on the scene from the start, and if you have been actively interested all along.

The most important process is the development of a warm, concerned, loving relationship between baby and parents. It is the basis for future relationships through which your child will be gradually civilized and 'disciplined' largely by parental example. Likewise the most effective sex education comes from being nurtured by loving, communicating, mature, and flexible parents. It is not what you *do* as parents that counts, but what you *are* as individuals and as a couple. Sometimes in the first few months you can be driven almost to distraction and may have the impulse to throw your baby out of the window or to strike her. Fortunately this almost always remains an impulse about which you should not feel guilty—it assails most parents. Above all, talk about it and the way you feel and always be prepared to listen to your wife's problems. If this becomes more than a passing impulse you should contact and talk to your GP about it.

Don't neglect your relationship with your wife. In the midst of the hurly-burly, anxiety, and general strife of coping with your baby and his needs, it is easy to neglect your interpersonal relationship. Try to put some time aside for yourselves to relax together. Take all reasonable opportunities to express your affection for each other. Touching, kissing, and cuddling don't take too much time in your busy schedule. Above all try *to maintain your sense of humour*—it is your most valuable asset.

Resuming sexual intercourse after the baby is born is an important part of your sexual relationship and one which will no doubt concern you both. You will have continued to kiss and cuddle, touch and comfort each other. Sometimes a woman may reject these advances if she fears they are an approach for intercourse too soon after the birth. Make sure you talk so that you understand each other. Simple physical factors such as stitches or the effects of a caesarean operation have to be taken into account. Remember also that you and your wife may have an exaggerated mental picture or fantasies about what has happened down there. If there has been a forceps delivery, or talk of tears inside or outside, or an episiotomy, then you and your wife may have lingering fears of damage. If doubts persist, go along with your wife for her post-delivery check-up (post-natal examination) or to your doctor at some other time and ask the questions—write them down beforehand if you like.

The healing process can be remarkably rapid and complete, and it's a good idea to at least encourage your wife to have a look with a mirror after a couple of weeks—she should be pleasantly surprised by the result. Delivery stretches the vagina and its opening, but the tissues have a surprising ability to return to their former size and shape. The levator muscles (the big muscles of the pelvic floor on either side of the vaginal opening) need to be especially re-trained to regain their tone. As well as being important in gripping during intercourse these muscles help to protect against developing prolapse later in life. Your wife should have been taught to strengthen these muscles by trying to pull the front and back passage up inside her as tightly as she can. She can do this while washing the dishes, while sitting on a chair, or at any other time. She must concentrate on pulling up as hard as she can for a few moments, hopefully without at the same time screwing up her face! She must do this at least 60 times each day for the first few months—muscles aren't strengthened easily. You can remind her and perhaps do other exercises with her.

The best time to resume sexual intercourse is when you as a couple feel you want to, provided you take note of some of the guidelines which I shall mention. It is not a good idea to put it off for too long, especially if there have been stitches for a tear or an episiotomy—the healing scar tissue can shrink too much so that the vagina narrows excessively. For most couples the time is between three and eight weeks after delivery.

There are a number of factors to consider. The blood-stained fluid which comes from the womb after the birth (the lochia) persists for three to six weeks, but may go on a little longer. It can be unpleasant but isn't usually too much of a problem. If your wife is breast-feeding the release of milk during sexual stimulation may be more than you have bargained for. Especially if a feed is due, the milk can stream out and make a real mess. Be prepared and perhaps choose a time after a feed. The let-down reflex responsible for milk flow is due to the release of the hormone oxytocin either because of the baby crying and your wife thinking about feeding, or due to the baby suckling. The same hormone is released by your stimulating the breasts, nipples, or vagina. Sometimes when your wife is breast-feeding her first baby, her whole concentration is on the baby and it can seem as though intercourse is not appropriate. It is as though, quite apart from tiredness, she finds it difficult to fill the role of both mother and lover. These are things you can talk about.

The possibility of introducing infection passes fairly soon, and certainly by two weeks. If there have been stitches for a tear or episiotomy, sexual intercourse would not cause the wound to break down after two weeks, but there is likely still to be considerable tenderness and soreness. Most don't start intercourse in these circumstances until after the postnatal examination which is usually at six weeks. However, no harm will be done provided you are gentle and your wife knows you will stop if she finds it too painful. At the post-natal examination we sometimes find granulation tissue (proud flesh) at the site of wound healing—this is the underlying connecting tissue which has grown too vigorously and sticks through the surface. It can be responsible for tenderness and pain. It is easily settled by coagulation with a touch from a stick of silver nitrate. Occasionally this stings a little. Complete healing then usually soon follows.

The vagina can heal with too much narrowing. This is not usually because stitching has been too tight but more a characteristic of that person's healing processes. It is interesting that I sometimes find this excess narrowing when I examine a woman who has been delivered by caesarean section without a preceding labour. The normal shrinking of the vagina to the pre-pregnancy dimensions is remarkably accurate but, as in these women, can err on the side of excessive narrowing. If this is a problem we explain to your wife how to pass two and then three fingers progressively further into the vagina each day to stretch it.

Sometimes the back edge of the vaginal opening can be drawn up to form a tender, thin ridge as it heals. This needs ironing out each day by your wife using her fingers and a little lubricant.

The final thing we find very occasionally is evidence of continuing inflammation in the pelvis. It is one of the causes of pain deep inside during intercourse. It may need treating with antibiotics.

It is a good idea to use a little lubricant such as KY jelly for the first few times as your wife's natural lubricating fluid may be a little slow in coming, and she may be too dry. Sometimes this can be because the normal balance of hormones has not returned, but more likely that tenderness and anxiety are interfering with the normal subconscious production of fluid which is a part of sexual arousal. The anxiety can be about possible pain or damage by intercourse, but can be part of an anxiety about having been damaged or changed in some way 'down there'. There is consequent anxiety about being able to function normally sexually. Don't forget that you too may not be as relaxed as usual and can have many of the same anxieties as your wife. Your anxiety communicates to your wife as, of course, your reactions in sexual intercourse are interdependent. If lubrication is insufficient then tenderness and pain are more likely and a vicious cycle can be set in motion. Above all remember that this is a time for patience and tender feelings, to allow more time for normal arousal.

Contraception should not be forgotten—full breast-feeding does greatly reduce the chances of conception but not absolutely reliably. Ovulation and therefore pregnancy can occur between four and five weeks after pregnancy—especially in those not breast-feeding.

One problem which is not often discussed is the effect on you of having been present at the birth as far as intercourse is concerned. So much depends on what happened and on your memories and feelings about the whole experience. If your wife has had a long and difficult labour and perhaps forceps delivery, which she has for one reason or another found traumatic, then inevitably there will be some effect on you. You may feel guilty about having been responsible for her having to go through the ordeal. You may in any case be concerned that your wife has some continuing damage or deformity as a result of the birth, a forceps delivery, a tear or an episiotomy. This will have been reinforced if you have seen the bleeding and, say, a tear before it has been stitched, so that you may have an exaggerated image of what is now wrong. You may be concerned that intercourse could interfere with healing, or do some further damage. Any anxiety you may have about resuming intercourse is likely to be reinforced if your wife is also anxious or withdraws and feels pain if you touch her there. You may both be further put off if intercourse does prove painful when it is attempted. There are undoubtedly some men who have been at the birth who later have difficulty in relating sexually to their wives. Their memories of what happened and images of what they think has happened can be very off-putting. Whether problems in resuming intercourse are more or less common in men who have been at delivery is not really known. Sometimes not being there can give rise to more exaggerated images-—the unknown can be worse than reality.

One thing is quite certain. Talking together about your anxieties and feelings about sex after the baby can help resolve them. If you or your wife have a continuing emotional or physical problem, don't hesitate to go and see your family doctor. He will be able to help you or will refer you to someone who can.

20 *Special situations*

Miscarriage

This is the loss of a pregnancy at a stage before the fetus is born alive. Most miscarriages occur in the first three months of pregnancy, and most of these are due to a chance error in the early development of the fertilized egg. The defect may be in the fetus, the placenta, or in the way the placenta is attached to the wall of the uterus. Occasionally, miscarriage occurs because of an acute or chronic illness of the mother. Some later miscarriages occur because there is a structural problem with the uterus, making it unable to retain the pregnancy satisfactorily until the fetus is adequately grown.

When careful research work has been carried out, an abnormality can be demonstrated in about 60 per cent of miscarriages, but it is very unusual for the defect to be familial or inherited—in other words, those women who have had one or two miscarriages and who haven't a predisposing illness start the next pregnancy with almost the normal chance of miscarrying, variously estimated as between 10–20 per cent. If the uterus has an abnormal shape which caused a late miscarriage the first time, it usually maintains the pregnancy considerably longer the second time. Those in whom this problem recurs may occasionally be candidates for an operation to correct the shape of the uterus.

Sometimes the cervix is too soft to resist the normal contractions of pregnancy and it opens too soon. This may be because the woman just happens to have rather soft, stretchy tissue. More often it is because the cervix has become damaged by a previous miscarriage, termination of pregnancy, normal or forceps delivery, or because an operation involving the cervix has damaged it. This is called the imcompetent cervix syndrome. It can be responsible for spontaneous abortion between 16–28 weeks or for premature delivery. It is treated by putting nylon or dacron tape or thick stitches (called Shirodkar or MacDonald stitches) around the top of the cervix under general anaesthesia. This is believed to make it stronger or more resistant. The tape is taken out two weeks before delivery is expected, or if labour develops early. Taking the stitch out is usually easy and doesn't need anaesthesia. Although the stitch can be put in before pregnancy, most agree that between 14–16 weeks is the best time if the problem is anticipated. Sometimes it is put in later when the cervix is found to be shortening and opening more than it should.

You and your wife will of course be sad and disappointed if a wanted pregnancy miscarries. Your reactions and emotions are similar, but less developed, than those described for stillbirth. You will both wonder why it has happened to you and will try to find a concrete explanation. This is usually lacking. It is hard to accept miscarriage as a manifestation of the inefficiency of natural processes. You, and particularly your wife, need to mourn this loss—to be able to talk about it over and over and to be upset or cry if you feel like it. Like stillbirth, miscarriage isn't something you can dismiss by 'snapping out of it' or 'looking to the future'. Time and appropriate sympathy and understanding are the essential requirements. Naturally the further the pregnancy has continued, the more your feeling of attachment has developed and the more you will feel the loss.

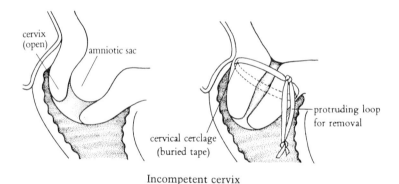

Incompetent cervix

There is no appropriate interval before another pregnancy, but it is usually better to allow a few months to collect yourselves emotionally. Sexual intercourse may be resumed according to your inclination, after allowing a few days for the lining of the womb to heal. Conception can occur as early as three weeks after a miscarriage if you don't use contraception.

As the placenta tends to separate incompletely from the uterine wall at miscarriage, it is usual to have an anaesthetic and a curettage (womb scrape) if the bleeding does not settle quickly. Ocassional women bleed heavily and need in addition a blood transfusion. Even though miscarriage can be so very upsetting, it is worth reiterating that you have an 80 per cent chance of success next time. It is very rare for spontaneous abortion to cause permanent damage (other than the incompetent cervix syndrome described above, which is occasionally caused by late miscarriage).

Threatened abortion (threatened miscarriage)

This is bleeding from the uterus during the first 28 weeks of pregnancy. There are several possible mechanisms:

1   It may be the start of miscarriage as the placenta separates from the wall of the uterus. The bleeding is not the primary cause, which is usually a defect in the development of the fetus or placenta. If miscarriage is going to occur, it usually does so within 48 hours. It is almost certain to occur if there is a significant amount of associated pain like period pains. The miscarriage can occur several days or even weeks after the bleeding has occurred. Sometimes the pregnancy dies but is retained within the uterus as 'missed abortion'. Without intervention this missed abortion usually miscarries sooner or later, but I have known of retention for more than a year.

2   The primary problem may be a break in a blood vessel supplying the placenta. The bleeding may be from the edge of the placenta in which case sooner or later the bleeding stops and the pregnancy continues. There are about 130 arteries supplying the placenta so that the loss of one does little to reduce the reserve of placental function. If the bleeding vessel is a little away from the edge, more placenta may separate, so that whenever there has been bleeding we suspect the reserve of placental function to be diminished. We don't therefore like the pregnancy to pass the expected date of delivery unless we have very good surveillance routines. Those who have bled in this way are more likely to miscarry later in the pregnancy or to have an early (premature) delivery.
  If the broken blood vessel is near the centre of the placenta and has bled extensively them miscarriage occurs. Some women bleed off and on through most of the pregnancy and still produce a fully grown, normal baby—in them, there is repeated leaking from near the edge of the placenta. These women are even more likely to miscarry or deliver early, and a few of them prove to have a placenta praevia (*see* p. 171). It is reassuring to note that, in these women who bleed and then continue the pregnancy, there is no greater chance that the baby will have a developmental abnormality than if the bleeding hadn't occurred. The bleeding is from your wife's blood vessel and not from the fetal vessels of the placenta: the fetus itself isn't bleeding.

3   The pregnancy may be ectopic (usually in one of the fallopian tubes instead of in the uterus). The bleeding in this case is from the tube or

from the lining of the uterus if the pregnancy in the tube isn't prospering. Women with an ectopic pregnancy usually have in addition a sharp lower abdominal pain. One in 200 pregnancies is ectopic. A laparoscopy (*see* p. 200) may be required to make the diagnosis.

4   The placenta may have undergone a change to a collection of fluid-filled sacs called a hydatidiform mole, which means that the fetus has died. This benign condition occurs once in 2000 pregnancies in Western Europeans.

5   The bleeding may be from some problem in the cervix or vagina and not actually from the uterine lining.

6   The woman may not be pregnant but simply has a delayed period.

With any threatened abortion, the first step is to diagnose the mechanism of the bleeding. This is done in the usual way by taking a history and then examining your wife's abdomen. Her pelvis is next examined with a speculum (small metal instrument used to inspect the cervix and walls of the vagina) and with two fingers feeling through the vagina. I must emphasize that if the pregnancy is normal, and in the uterus, these examinations won't harm it. The pregnancy is inside the thick-walled uterus and out of harm's way. The information obtained by these examinations may be vital for your wife's welfare.

In any woman who has had bleeding, further investigations in the form of an ultrasound scan may be needed. This checks on whether or not there is a pregnancy in the uterus and whether or not the fetus is alive and thriving. A laparoscopy under general anaesthetic may be needed to confirm or exclude a suspected ectopic pregnancy.

Treatment depends on what is found—if the miscarriage is incomplete, or there is a missed abortion or a hydatidiform mole, then the uterus is emptied under general anaesthesia. If the pregnancy seems normal then there is perhaps a case for the woman resting at least until the bleeding stops if the pregnancy has gone beyond four months. It is not clear that this helps, but it might.

If the woman is less than four months pregnant then it is almost certainly a waste of time for her to rest. The fate of the pregnancy is already sealed and is not influenced by rest. I say this on theoretical grounds (I cannot imagine any mechanism by which rest could help) and also because practical surveys have not demonstrated benefit. None of the pills and potions or hormone preparations sometimes prescribed have been shown to be of value. If the pregnancy is destined to miscarry, it miscarries, and if to continue then it will.

Sexual intercourse has not been shown to be harmful when a pregnancy has threatened to miscarry, but we know that the uterus contracts strongly at orgasm so perhaps sex should be avoided for at least a week after the bleeding has stopped. Many couples decide not to indulge until after the fourth month.

Anaemia

Anaemia signifies a lowered concentration in the blood of the red pigment haemoglobin and can arise in several ways. It may follow haemorrhage because the plasma (the colourless fluid part of the blood) is restored within 48 hours whereas the red blood cells which contain haemoglobin are regenerated more slowly over several weeks; in pregnancy the fetus may utilize the important building materials iron and folic acid, and if these are in limited supply the mother becomes anaemic—hence the frequent need to give these as supplements; a disorder causing anaemia may be detected by routine testing in pregnancy—examples are thalassaemia of Mediterranean and Asian peoples and sickle cell disorders of Africans. Normally in pregnancy the total blood volume rises but the increase of plasma is relatively greater, hence the normal haemoglobin concentration of pregnancy of 11 grams per decilitre is lower than the level of the non-pregnant 12–13 g/dl. It is not uncommon to have what is called the physiological or normal anaemia of pregnancy where the level is below 11 g/dl simply because the relative rise of plasma has been even greater than usual. The management of anaemia is to try to identify the underlying cause before giving treatment as appropriate. The commonest is iron deficiency treated with iron tablets or injections.

Toxaemia of pregnancy (pre-eclampsia)

This problem is diagnosed when the examination by the midwife or doctor finds two of the following three features arising in the second half of pregnancy: raised blood pressure (hypertension); protein in the urine (proteinuria); and generalized swelling (first noted as a rapid rise in body weight). One of the main reasons for regular antenatal checks in late pregnancy is to detect this condition at a relatively early stage. When the woman actually complains of symptoms (with the exception of swelling), the condition has reached a relatively late stage. The other symptoms which might appear then are headache, drowsiness, or flashing and spots before the eyes.

The cause of toxaemia is unknown, although it is more likely to develop in a first pregnancy or in a fourth or a later pregnancy, in a twin pregnancy, or when the woman has diabetes, chronic kidney disease or pre-existing raised blood pressure. The term toxaemia is a misnomer, as toxins or poisons are not known to be involved. It is not easy to decide whether a raised blood pressure recorded at a clinic is a significant indication of impending toxaemia. It is normal for anyone's blood pressure to rise when she is anxious, worried, or annoyed, or when she has been rushing around. If your wife has rushed to the clinic or been worried or annoyed by something said to her, or even if she is concerned that the blood pressure may be raised, then it is likely it will be. It is therefore checked again at an interval after her arrival. If it is still significantly raised, she is admitted to hospital for a day or two, or spends an afternoon or morning having her blood pressure regularly checked in an assessment clinic. If her blood pressure returns to normal, then the earlier readings can be ignored. It is only when the blood pressure is raised when your wife is resting and relaxed that it may be important. Some women are more relaxed at the GP's surgery or if a midwife calls at home, so a hospital clinic may ask for this form of assessment.

The main treatment is bedrest, usually in hospital, and sedatives and anti-convulsant drugs are given if the condition is moderate or severe. Toxaemia reduces the efficiency of the placenta in supplying the fetus with oxygen and nutrients. He may not grow at the normal rate and if left could even die in the womb. Most cases of toxaemia arise near the end of pregnancy, but if it should start early, blood-pressure-reducing drugs may be given to try to control the blood pressure until the baby is mature enough to have a good chance of survival.

The most effective treatment is to deliver the baby. The doctor has to decide when this should be and whether it should be vaginally after induction of labour or by caesarean section. When the woman is in labour she needs very good pain relief, for trying to cope with pain may further elevate the blood pressure to a dangerous level. An epidural is often the best method as it not only controls the pain, but has a direct effect in lowering the blood pressure by relaxing blood vessels.

The time to effect delivery is when the disadvantages to the baby of being born early and any associated disadvantages to the mother are judged to be less than the chances of the baby dying of undernourishment if he stays inside. Fortunately, this is usually not a knife-edge decision, and the baby almost always does well. Occasionally the toxaemia worsens to the extent that the baby has to be delivered to protect his mother's health. This is not usually against the baby's interests as severe toxaemia is also a menace for him, but he may have to be delivered when very immature and therefore may not survive.

In these more severe cases the mother, if undelivered, is in danger of having fits, severe haemorrhage from the placenta, or other complications. She is kept quite sleepy with sedatives for one to two days after delivery. Once over this acute phase, recovery is complete without persisting detriment to health. If there is no underlying cause for the toxaemia there is only one chance in five that it will recur with the next pregnancy, and even then it is usually less severe.

Rhesus factor problems

The term 'rhesus factor' was coined after it was shown that the blood of 85 per cent of western Europeans agglutinated (stuck together) when the serum of the rhesus monkey was added. The factor on the surface of red blood cells responsible for this interaction was called the rhesus factor—the 85 per cent of women who have this factor on the surface of their red cells are called rhesus positive and the 15 per cent without it are called rhesus negative. The proportions are different in different racial groups.

If a rhesus-positive man marries a rhesus-negative woman the first resulting fetus may be rhesus positive. When the placenta separates at delivery an average of half a millilitre of baby's blood with the rhesus marker protein can enter the mother's bloodstream. As the mother's system recognizes this protein as different and foreign, it slowly develops the ability to form antibodies against the rhesus factor so that it is ready for the next invasion. Sometimes this sensitization does not occur under these circumstances. Also the first rhesus-positive fetus may not be conceived until a later pregnancy.

In the pregnancy following the one causing sensitization the few fetal cells which cross into the mother's own circulation during pregnancy are sufficient to stimulate the rapid production of antibody in the now sensitized system. These antibodies then, in the second half of pregnancy, cross back into the fetus's circulation where they agglutinate and destroy some of the fetal red cells. If the rate of destruction exceeds the rate of production the fetus becomes anaemic.

The effects on the fetus are of varying severity. The least severe problem is that the fetus is apparently normal at birth but becomes jaundiced within a day or two, and may need to have her blood changed by exchange transfusion to eliminate the excess yellow pigment. Secondly, the baby may be quite anaemic at birth and need immediate exchange of blood to prevent some anaemia and to treat the jaundice. Alternatively the fetus may be so severely affected that it dies in the uterus unless given an intra-uterine transfusion. Sometimes the baby is rescued by delivering it a little early.

This process is detected during pregnancy if the rhesus antibodies are found in one of the mother's blood samples taken during pregnancy. A more exact assessment of the severity of the process is made by serially analysing samples of the liquor taken by amniocentesis.

The sensitizing process can be almost certainly prevented if rhesus negative women are given, within 48–72 hours of delivery or one of these episodes, an intramuscular injection of gammaglobulin containing rhesus antibody. This antibody destroys the fetal cells in the mother's circulation before they are able to stimulate the antibody-producing mechanism in the mother.

The whole rhesus process is rather more complicated than I have described. For instance rhesus-positive women can develop antibodies to some of the lesser rhesus sub-groups. A rhesus-positive woman is positive to the main rhesus-group factor but may be negative to other minor rhesus factors of lesser importance. All women, whether rhesus positive or negative, need to have their blood tested during pregnancy for the various lesser rhesus antibodies.

Breech presentation

This term is used when the breech, or bottom end, of the fetus is lowermost in the mother's abdomen. By the end of pregnancy more than 96 per cent of babies present by the head end, the breech end with the legs and feet being bulkier than the head and more easily accommodated in the wider upper end of the uterus. With the amount of liquor relative to the size of the fetus diminishing as term (the due date) approaches, the fetus has less opportunity for turning head over heels. He does it often before this stage and sometimes he does it once too often and gets caught with his bottom down (presenting). Sometimes the fetus continues to change presentation right up to the time of labour and, rarely, even in early labour. Many obstetricians try to turn the fetus to a head presentation when a breech is found presenting after 34 weeks. It is a relatively simple process. After putting talcum powder on your wife's abdomen, to allow the doctor's fingers to slip, pressure is applied to the head and breech ends in opposite directions to encourage the fetus to kick herself around. This requires your wife to relax her abdominal wall but is at most only mildly uncomfortable, as we have no wish to apply any extra pressure. Not all doctors agree that we should attempt to turn the fetus in this way, and in some circumstances the manoeuvre is contra-indicated. Sometimes the attempt to turn the fetus is unsuccessful and we are then faced with breech delivery.

We then carry out a careful assessment of the size of the baby (sometimes with the aid of ultrasound); of the baby's posture (with an X-ray); and of the size and shape of the bony part of your wife's birth canal by internal examination with the fingers and by X-ray. If the baby is not

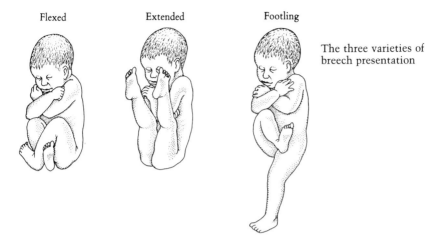

Flexed   Extended   Footling

The three varieties of
breech presentation

large, has a favourable posture, and there is a roomy pelvis then breech
vaginal delivery is preferable. Otherwise caesarean delivery is advised.

Breech labour is essentially the same as that with the head presenting.
If it doesn't go along well then we change to caesarean.

For the actual delivery your wife lies with the lower half of the
delivery bed removed and her legs relaxed and supported in stirrups.

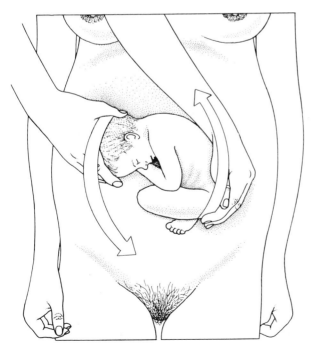

Turning (version) of
the fetus to a head
presentation

This is essential for the safety of her baby so that she can be guided safely towards the floor during the early part of the delivery and any other manipulation which may become necessary can be carried out quickly and efficiently. Your wife may be surprised how comfortable she finds this position if she can have the confidence to let her legs go loose. She can push very effectively by holding the poles to which the stirrups are attached.

If an epidural hasn't been used, the opening of the vagina is numbed with local anaesthetic (pudendal block or local infiltration.) This is an essential precaution for the safety of the baby, as it reduces the chances of the mother suddenly feeling pain during the delivery. She is thus more able to keep her buttocks reasonably still. Otherwise your wife doesn't feel anything different in breech labour and delivery from the feelings described for a head-first delivery. She pushes down very effectively in second stage by holding on to the stirrups and pants slowly as the baby emerges, or gives an extra push if asked to. The baby usually looks very blue at delivery. It is usual to have an episiotomy to relieve pressure on the head and most doctors prefer to use forceps to make sure the head is delivered slowly and therefore safely. The forceps make sure the head doesn't pop out quickly. A less favoured alternative is to ease the baby's head out with hands placed on it in the vagina.

Twins

Twins may be univolar (from one egg, identical), or binovular (from two separate eggs, fraternal, i.e. no more alike than other siblings). The rate of uniovular twinning is constant throughout the world at 3 per 1000, and seems to be a chance phenomen without any inherited tendency. The fertilized egg starts to divide but, instead of going on to form only one fetus, splits into two identical halves, each of which forms an individual.

With fraternal twins the woman seems to have a tendency towards double ovulation (release of two eggs at a time) inherited from her family. Each egg is separately fetilized. Binovular twinning occurs 11–12 times per 1000 in Britain, but as high as 46 per 1000 deliveries in the Yoruba tribe of Nigeria, and only 1 per 1000 in Japan. A woman in Britain with a set of binovular twins has 1 chance in 8 of a repetition next time.

Twins are suspected when the uterus is larger than expected for the stage of pregnancy; when the usual symptoms of pregnancy are exaggerated; or because you or your wife thinks there is a good chance of twins—perhaps because of a family history on your wife's side. The diagnosis is most easily confirmed by ultrasound and should be detected

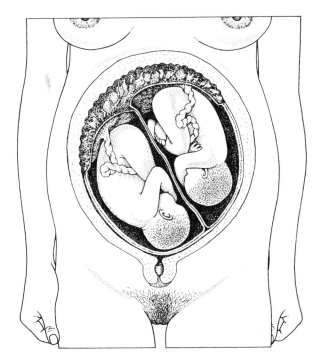

Most identicals

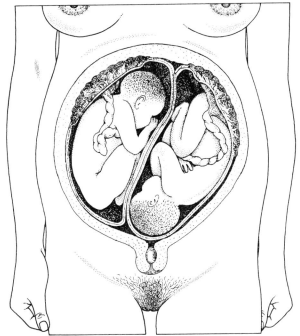

Most non-identicals

at15–16 weeks if this is the time for a routine ultrasound examination. It is now rare for twins to be detected only by counting them as they deliver!

Twin pregnancy is much the same as with a single pregnancy but everything is accentuated and there is a greater strain on the body's resources. More rest is indicated in late pregnancy and more frequent antenatal checks are in order. Anaemia, toxaemia, and premature delivery are more likely. The placenta is more likely to start to fail early.

Towards the end of the pregnancy, when your wife seems doubly awkward, tires more easily, and is concerned about her appearance, your reassurance, support, and love will be even more important. Concentrate on the positive aspects of the twin pregnancy and the excitement and challenge you will both be meeting. Buy a book on twins as there are many fascinating aspects you can explore together.

Many labours start early, but induction because of evidence of unsatisfactory placental function or raised blood pressure is more likely. When considering the induction of a prolonged pregnancy, 38 weeks is taken as the reference point for term rather than the 40 weeks of a singleton pregnancy. Most doctors aren't keen on a twin pregnancy going past 40 weeks because of the unreliability of placental function.

It is quite possible that the second fetus may turn or need to be delivered as a breech regardless of whether or not they are both presenting by the head at the start of labour. The decision between vaginal delivery and caesarean section for the twins is made in the same way as for a breech presentation.

During labour both babies should be carefully monitored. Apart from this and favouring epidural analgesia a bit more strongly than usually, labour proceeds and is managed much as for singleton pregnancy. For the second stage we prefer to have a drip running into a vein so that it is ready if needed to stimulate the uterus to deliver the second twin if it takes too long to re-start contracting after the first delivery.

Delivery position and local anaesthesia are as for breech delivery. When the first baby has delivered there should be particular concern for the second baby, as there can be special problems in making sure he is delivered safely.

We first check that his heart rate and rhythm are satisfactory and then check his lie (his long axis in relation to his mother's long axis). If he is lying transversely (crosswise), he is manipulated into an up-and-down (preferably head first) lie by manipulation through the abdominal wall. The bag of membranes in which he is contained is then broken and the presence of prolapsed cord or bleeding from the uterus excluded. If the uterus doesn't soon spring into action after its well-deserved rest following the first delivery, an oxytocin solution is run through the drip to stimulate the uterus. We prefer the interval between delivery of the babies to be not longer than 15 minutes as after that the likelihood of oxygen lack through placental failure increases. The second baby has no first stage as the cervix has already been fully stretched and second stage is short as the lower birth canal is soft and offers only minimal resistance. Sometimes if he persists in lying crosswise or needs rapid delivery it can be necessary to turn the baby to breech presentation by internal manipulation under epidural or general anaesthesia (internal version). Although all this may sound rather complicated, it is quite straightforward for the experienced obstetrician and we expect to end up with two healthy babies and two happy parents.

Many mothers successfully breast-feed twins and you will be given special advice about this.

Antepartum haemorrhage

Antepartum haemorrhage is officially vaginal bleeding after the 28th week of pregnancy, though bleeding a little earlier has similar significance. As with threatened abortion the first priority is to establish the cause. Our chief concern is that there may be a placenta praevia. This is where the placenta is partially attached to the lower segment instead of being entirely on the upper part (body) of the uterus. Placenta praevia occurs only in 1 in 200 pregnancies but it is very important as serious haemorrhage may occur—usually some time after the first warning bleed. You will remember that the lower segment is the part of the uterus which stretches and thins out towards the end of pregnancy. It is easy to see the mechanism of bleeding. As the area of the wall to which the placenta is attached increases, a blood vessel passing through the wall to the placenta is bound to break. Sometimes this isn't until after labour has started.

Any bleeding in the last 16 weeks of pregnancy should be reported without delay. The doctor may then check its origin, if it is a small amount, by looking with a small instrument as your wife lies on her side. Unless there is an obvious bleeding point, an ultrasound examination to check the site of placental attachment is the next step. If this shows a placenta praevia, your wife is likely to be kept in hospital with blood cross-matched in readiness for a transfusion should more serious bleeding start. She would then be delivered by caesarean. If the pregnancy continues until 37 weeks are completed, then delivery would be advised either by caesarean section or after induction of labour if further examination shows only a minor encroachment of the placenta on to the lower segment. To wait longer when the fetus has attained this adequate maturity is to run a now unwarranted risk of further bleeding.

If the ultrasound excludes placenta praevia your wife would come home again with the advice to have extra rest, which may reduce the risk of delivering early.

Occasionally serious bleeding from a normally situated placenta can start for no apparent reason (called *accidental haemorrhage* as it was once thought to be mostly associated with accidents). Sometimes the baby can be saved by speedy resort to caesarean section but more often he dies because too much of the placenta has separated. This problem is sometimes the sequel of toxaemia of pregnancy and anticipating it is

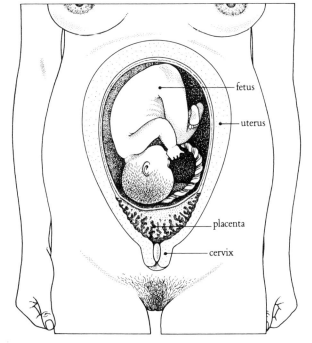

Placenta praevia

fetus

uterus

placenta

cervix

one of the reasons for early delivery if the raised blood pressure persists too long.

Small babies

Seven to ten per cent of all babies are born small either because they are growing too slowly (small for dates) or because they are born too early (pre-term).

Babies are small for dates usually because the placenta is under-performing (placental insufficiency), but sometimes because the fetus has developed abnormally. One tries to distinguish these two problems when it is suspected that a fetus is undergrown, but is not always successful. Ultrasound and sometimes X-rays are the main investigations at this stage.

Placental insufficiency cannot be treated effectively, even though efforts have been made to introduce treatments such as intravenous infusions of various nutrients to the mother. Sometimes the mother has a definite predisposing condition such as high blood pressure and controlling it can perhaps help a little. Extra bedrest for the mother is our traditional treatment—it may help but we are not sure. The most important aspect of management of placental insufficiency is close observation of the fetus's condition with a view to early delivery. (This is sometimes by caesarean if it is thought that vaginal delivery may be hazardous.) Delivery is effected when the dangers of early delivery are less than the danger of the baby dying in the uterus. Once delivered and given expert paediatric and nursing care, these babies do well and gradually catch up growth though they sometimes remain on the small side. Your wife, and to some extent you, will be disappointed that you haven't produced a well-grown baby. There is nothing wrong with being disappointed, but you will realize that she too is a miracle of creation and you will come to love and cherish her.

Pre-term babies are of a weight appropriate to their age, but have arrived early (before 37 completed weeks). Most times they are quite normal except for the general immaturity of their organs and body systems and controls. They are rather more likely than term babies to have a developmental abnormality. They have arrived before they are fully ready to live outside the womb, and usually need extra paediatric and nursing care. If they are very small they need all the resources of an intensive baby care unit. Sometimes they have arrived early because we, the obstetricians, have rescued them from a deteriorating intra-uterine environment; sometimes the cause of the early labour is evident in early rupture of the membranes or partial separation of the placenta; but most times the reason for this early arrival is not clear.

Your wife should be admitted to the labour ward whenever you think premature labour is threatening. We can sometimes postpone the time of delivery by running into a vein one of a series of substances which reduce uterine contractions.

Delivery of small babies needs extra care, and episiotomy to reduce pressure on the small head is routine.

Induction of labour

Starting labour artificially instead of waiting for the natural onset is usually resorted to when any risks or problems associated with induc-

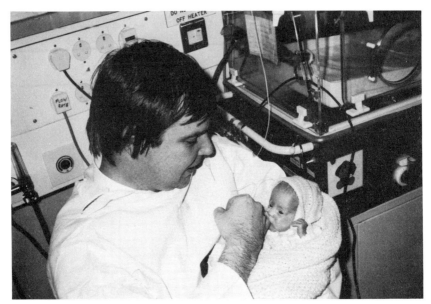

Danielle, born weighing 1 lb 14 oz, is now doing well. She sucks her father's finger during his visit

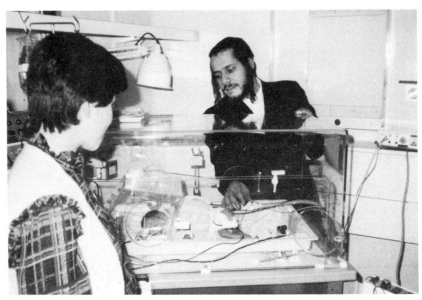

This young man, who will soon be big enough for transfer back to his local district hospital, fascinates his parents

tion are less than the risks to the fetus of staying longer in the uterus. Most often there is reason to believe that the function of the placenta in supplying nourishment and oxygen to the fetus is declining to a dangerous level. Sometimes labour is induced principally because of a hazard to the mother such as fairly severe toxaemia of pregnancy, or haemorrhage from the placenta.

A principal disadvantage of induction is the occasional failure of the uterus to respond adequately to the stimulus and go into normal labour. If the method for attempting induction has involved breaking of the

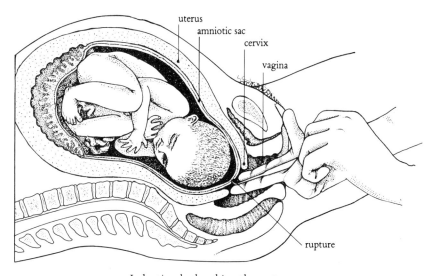

Induction by breaking the waters

waters, then failure to respond means delivery by caesarean section with its relative disadvantages. We would therefore not proceed to breaking the waters unless we were fairly certain of a good uterine response. There are, however, occasions when the risks of not inducing are sufficient to justify this small risk of precipitating a need for caesarean. We assess the likelihood of a good response by considering several aspects: the stage of pregnancy—the nearer term the better the response; whether or not the head is engaged in the pelvis—engaged responds better than non-engaged; the state of opening (ripeness) of the cervix—the most favourable is a cervix which has taken up (lost its length) started to open (dilate), is relatively soft, and is relatively thin.

Sometimes in a multigravid patient labour starts fairly abruptly following the various methods of induction, but usually it develops gradually, as when labour starts spontaneously, although the preliminary phase is often relatively short with either first or subsequent babies.

There are several methods of inducing labour, which may be used singly or in combination. One may, under some circumstances, desist from attempting induction when a simple technique has failed, rather than proceed with, say, rupture of the membranes if the need for induction isn't sufficiently strong to justify this. Some methods have been completely superseded by better ones. A few of the older methods are still used by some doctors. The three commonly used methods of induction are:

1   *Prostaglandin* in the form of tablets, pessaries, or gel put high in the vagina near the cervix, the dose being repeated several times if necessary. Even if labour doesn't start, the cervix may become more favourable so that the membranes may be broken in the expectation that labour will follow.

2   *Breaking (rupturing) of the membranes* (surgical induction of labour). This is the most effective method and also the most appropriate in many circumstances. Allowing escape of liquor alters the pressure relationships in the uterus and triggers labour. As the membranes have no nerves in them, their breaking is painless but in order to reach the membranes the cervix may need to be stretched a little. This can be quite painful but for most is just uncomfortable. Some women can find the whole process difficult to cope with, especially if it proves painful; if they are not able to relax well; or if the doctor is not gentle and adequately reassuring. Sometimes a pain-relieving injection will be given beforehand, or can be given at the time if the process proves unexpectedly painful. If an epidural is to be given for pain relief in labour it may be started before the membrane rupture. Very rarely a brief general anaesthetic may be given for the induction.

3   *An oxytocin drip.* The naturally occurring substance, oxytocin, produced by the pituitary gland and concerned with uterine contractions in labour, has been made artificially. A solution of this is run into a vein and the rate is regulated to produce the strength and frequency of contractions one would expect in a normally progressing labour. This is, of course, persisted with only if the fetus isn't showing any evidence of becoming short of oxygen. The drip solution is not usually used until after the membranes have ruptured naturally, or have been ruptured artificially. If, after rupture of the membranes, the uterus does not within a few hours begin to contract satisfactorily, this additional stimulus is used. It is the same solution as that used for the augmentation of labour. If during the course of labour progress is so slow, once

the membranes have been ruptured, that it seems unlikely to be completed within 24 hours, and if there is no apparent reason for this slow progress, then the oxytocin solution is used to stimulate the uterus to contract normally (augmentation).

A decision for induction is often far from simple and can require considerable professional judgement. One of the problems of induction is that the baby when delivered may prove to be unexpectedly immature. Very careful assessment of the evidence for deciding the age of the fetus is always a must. This is now easier if a reliable ultrasound assessment of fetal age has been made earlier in pregnancy. There are differences of opinion about how much risk the fetus runs when the pregnancy continues more than two weeks after the expected date, when this is known with reasonable certainty. The risks of continuing the pregnancy can be minimized and therefore the need for induction at this stage reduced if there is a well-organized and efficient local facility for assessing how well the placenta is continuing to function. It is still unusual to have such good facilities. Where these facilities are not good then it may be safer to induce even low-risk patients after 42 weeks. The time of induction is brought foward when there are other risk factors such as raised blood pressure or relatively advanced maternal age.

What say can you have in this matter of induction? It is easy to have relatively strong views for or against induction, but not so easy to hold valid views. If reasonably possible, you should go along with your wife when induction is likely to be considered, so that you may both have the situation explained and have an opportunity to ask questions. It is usually wiser to be guided by your professional advisers, but above all don't hold inflexible views about induction as you could prejudice the future of your baby.

Despite the commonly projected impression that patients are against induction, I find in practice that I more often have to talk patients out of having an induction when they think pregnancy is going on too long! You would be surprised how often couples start out in in pregnancy with anti-induction ideas and then push for it when the onset of labour is delayed. Often our reluctance is based on uncertainty about the age of the fetus. You, and especially your wife, may wish to have an induction partly because the last part of pregnancy can be a bit dreary. She may not sleep well, feel clumsy, heavy and tired, and may feel that the pregnancy will continue for ever.

Your underlying desire to have a fit, healthy baby and your anxiety that something may go wrong not surprisingly can get too much for you. Your friends and relatives don't help by ringing your wife each day to say, 'Are you still there?' 'Are you sure the labour shouldn't be induced?' etc. A few couples add two weeks in announcing the expected day to reduce this harrassment but I would counsel you against using this deception with your professional adviser!

It is worth remembering that an important objective of pregnancy is to have an alive, fit, healthy baby, and that this is very much in the minds of your doctor when he/she decides whether or not to advise induction.

Episiotomy

This is a minor procedure to enlarge the opening of the vagina when necessary just as the baby is about to deliver. It isn't made at the beginning of the second stage as some people think. The baby usually delivers with the next one to three contractions. After inserting local anaesthetic to numb the tissues, a short cut is made in the mid-line between the vagina and anus, or more commonly at an angle of about 45°

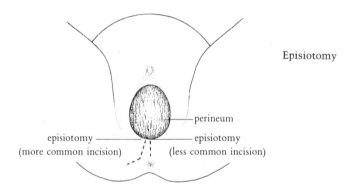

Episiotomy

perineum

episiotomy
(more common incision)

episiotomy
(less common incision)

starting in the midline and moving out towards a spot to the side of the anus. The actual cut is not painful. The woman gets a feeling of relief that there is suddenly more room. Sometimes you and your wife can hear the cut being made.

This cut is made either in the interests of the mother or the baby. In well over 70 per cent of women having a *first* baby without the aid of episiotomy, the tissues tear. Many of these tears don't matter too much but others are serious in the short, and sometimes in the longer, term. At delivery a woman feels the tear only as part of the strong, sometimes very painful, stretching process which allows the head to deliver. Delivery of the head gives some sense of relief which is completed when the whole baby is delivered. Sometimes when an episiotomy is indicated and isn't made when tough resistant tissues are being slowly stretched, then the final stretching just before the tissues tear can be excruciatingly painful.

The main problem with letting tissues tear is that one can't be sure how extensive the resulting tear will be. Its extent is evident only after the delivery. Extensive tears can bleed freely at the time and can occasionally be extremely difficult or even impossible to stitch satisfactorily, even under general anaesthetic. A woman who has had a bad tear can occasionally be left with less certain control over the escape of wind from the bowel, and sometimes the entrance to the vagina can become lax. Although it is difficult to get conclusive evidence, one certainly has the impression from experience that later troubles needing repair operations for prolapse are more likely when tissues have been over-stretched or torn. One problem with muscle is that in tearing it comes apart very irregularly. It is for these reasons that the straight cut of an episiotomy can be better for the woman when it appears that the tissues will tear. The earliest evidence that a tear is starting is the presence of bleeding as the head recedes a little between contractions. Most tears begin on the inside and even extensive tears may leave the outside skin intact.

Apart from protecting the mother's tissues in this way, episiotomies are made in the interests of the baby. The time just before delivery is easily the most dangerous for the baby because of possible inadequate oxygen supply. Even though there is sufficient reserve of supply for most babies to deliver safely, occasional babies can become acutely deprived of oxygen and even die at this stage without giving much warning to the attendants. The worry about oxygen lack is that the baby's brain depends on its supply of oxygen and its deprivation can produce permanent brain damage. Two factors contribute to this problem. When the baby is stretching the soft tissues at the opening (the perineum) a large proportion of the baby has already moved out of the uterus and the liquor has been lost. As the uterus contracts to accommodate this loss of volume, the blood flow to the placenta can be impaired with resulting diminution of the oxygen supply to the fetus.

The other factor is that the baby's lifeline, the umbilical cord, can become compressed or drawn tight, especially when it is around the baby's neck. This too can diminish oxygen supply to the baby by impairing blood flow in the cord. It is therefore good practice in the interests of the baby's brain to make an episiotomy if the baby's progress becomes slow in these final stages. An episiotomy is a very small price to pay for making more sure that the baby's brain functions remain intact. It is a minor procedure which occasionally averts a serious disaster. If there is evidence of the baby becoming too short of oxygen during the first stage of labour, we don't hesitate to advise a caesarean delivery. This is an altogether much bigger operation than the episiotomy sometimes indicated at delivery.

Episiotomy is almost always used at forceps delivery to protect the mother's tissues, and at breech delivery to protect the baby's head from pressure and the ensuing sudden release of pressure. If a premature baby is delivering, it is essential to protect it from being squashed by doing an episiotomy—the smaller the baby's head the more important this is. Episiotomies are also used for some medical conditions of the mother such as chest and heart problems, and when a previous delivery has been by caesarean. You can usually nurse the baby and hold your wife's hand while the repair is being done.

Methods of stitching a tear or an episiotomy are now much better so that stitches usually don't come through the surface and don't need removing. This means they cause less pain on movement. Some doctors still use removable stitches, especially for some types of tear. Pain is less than it used to be but some, especially if they also have trouble with 'piles', can have a lot of pain for a few days. Pain and soreness have usually almost completely disappeared in three to eight weeks, but a few are unlucky and have pain and tenderness for some months.

Your wife needs an internal examination about six weeks after delivery to make sure internal healing is complete. Many women are troubled by pain on first resuming intercourse, and in some it can be quite persistant. If this pain is a real problem or is still present in another one to two months encourage your wife to go back to her doctor. In the only good studies which have been carried out there has not been any difference in the overall incidence of such pain in those who have had episiotomies compared with those who have had tears. The lesser trouble after small tears is very much cancelled by the greater trouble after large and multiple tears. Because most times when at delivery a tear seems likely the midwife or doctor will not be able to predict with any certainty, whether the tear will be large or small it is wiser to opt for the controlled cut of an episiotomy.

There is no doubt that everyone would prefer *not* to have a tear or an episiotomy, but for the majority of *first* deliveries it is one or the other. This is really another example of a none-too-efficient mechanism of nature.

There has been a tradition of trying at almost all costs to keep the perineum intact at delivery. This is in my view very bad practice and sometimes dangerous. All women delivering need managing as individuals. How delivery is managed must depend on the circumstances at the time and on the skill and judgement of the deliverer if the best results are to be obtained. While recognizing that everyone prefers not to have any need for stitches, this must be within the bounds of safety for mother and baby. People are sometimes given the impression that they can choose at the time whether or not to have an episiotomy.

Any discussion or explanation needs to take place earlier, as usually a decision for episiotomy needs to be made rapidly just prior to delivery, and there is only time for a comment but not for a discussion. It is quite reasonable for your wife to say during such a prior discussion that she would prefer to have an episiotomy rather than run much risk of a bad tear, or to say she would be happy to run some extra risk of a bad tear in the hope of not needing stitches provided the baby is not at risk. Unfortunately it is rarely possible with a first delivery to be sure that if a tear occurs it will be small.

Forceps delivery

Obstetric forceps are designed to assist the delivery of your baby. Forceps delivery is performed for one of many reasons.

1   Commonly because progress in the second stage of labour has stopped or become very slow. This is either because the baby's head is fitting tightly or because the baby's head is the wrong way round (the back of the head not facing the front), or because the muscle of the womb has become tired and is not doing its job well. This is most often the case with a first labour. A woman can be pushing down very efficiently and yet be unable to make progress in the second stage because the uterus is just not pulling its weight. She must have the help of the uterus. Occasionally, it is the mother's pushing-down effort which is not so good. Another reason for slow progress can be that an apidural is being used for pain relief. The epidural can interfere with the strength of the woman's pushing effort and can, by excessively relaxing muscles of the pelvic floor, interfere with the normal rotation of the head.

2   Often because the baby is considered to be becoming so short of oxygen (fetal distress) that further delay in delivery could run the risk of the baby suffering brain damage, or even dying.

3   Sometimes because the mother has a problem such as high blood pressure or heart disease, so that pushing in the second stage could be dangerous for her health.

4   In breech delivery, to make sure that the rate of delivery of the head is slow and controlled to protect the baby's head from a sudden release in pressure.

Forceps deliveries are called low or outlet deliveries when the baby's head is starting to stretch the big muscles and soft tissues of the outlet of the pelvis, but is not far enough down that simply doing an episiotomy will allow the head to deliver.

Mid-forceps delivery is where the baby's head is a little higher. In some of these the head will need to be turned (rotated) so that the back of the head (occiput) comes around to the front before delivery can be completed. This is a rotational mid-forceps delivery.

The woman is given what pain relief she needs for the procedure. For low forceps the tissues at the opening are numbed with local anaesthetic or by putting the local anaesthetic a little higher to block the nerve taking sensation from the area of the opening (pudendal nerve block). The woman must push down strongly and let herself go while the doctor pulls down. The same measures are usually adequate for ordinary mid-forceps.

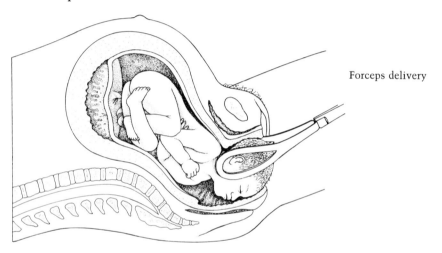

Forceps delivery

For rotational forceps deliveries an epidural or a general anaesthetic is required as the procedure would be too painful otherwise.

If an epidural block is already working for labour then it can be reinforced by sitting the woman up and giving a top-up dose to give good pain relief for all types of forceps delivery.

If these pain-relieving measures are not working well, or if your wife is not pushing down and letting herself go when pudendal nerve block only is being used, then the forceps delivery can be distressing and painful. These simpler measures are not sufficient and are better replaced with an epidural or general anaesthetic if a woman is distraught or not able to co-operate well.

An episiotomy is used to prevent tearing during delivery. (The line of the cut is numbed by the above measures.)

Forceps are now very safe for mother and baby. In earlier times when caesarean births were dangerous, every effort used to be made in the interests of the mother to make sure the baby was delivered by forceps.

This used at times to result in the baby being damaged by the forceps. The type of forceps delivery which could damage the baby has now quite rightly been replaced by caesarean delivery.

It is not unusual nowadays for you to be present at a forceps delivery if your wife is not having a general anaesthetic. The doctor, your wife, and you need to agree to this. You should not, however, feel any obligation about it—it is a very individual matter. Again I would advise that your interest and attention are with your wife at the top end. The forceps can seem rather large and frightening to you and there is not much to be said for you being too closely interested in the mechanics of the procedure at this stage. You can be of great comfort and encouragement to your wife and you can share in the joy and magic of your baby's arrival. Don't forget that there is bound to be some blood on the baby, but you will hardly notice it if you realize it is normal.

Both you and especially your wife may have said all through pregnancy that the last thing you want is a forceps delivery. Don't be surprized however, if at the time your wife is very relieved when the forceps are suggested. She is likely to say, 'Hurry up and get on with it' or: 'Do anything you like to get it over and done with!' You, in her situation, would react similarly. When the decision has been made to have a forceps delivery then the second stage can suddenly become quite unbearable. Your wife gives up and everyone has a feeling of impotence if there is any wait for an anaesthetist. Your wife should continue to bear down with contractions—this is the best way to relieve the pain low down in her abdomen which appears in second stage when a woman isn't pushing. You can encourage her.

If your wife has had forceps for a first delivery she is very unlikely to need forceps the second time. The uterus works much more effectively and the tissues which have been stretched once yield much more readily. Even if the second baby is in a posterior position it will turn and deliver much more readily then such a first baby would.

Ventouse (vacuum extractor) delivery

This is an alternative to forceps for assisting or speeding up delivery. It is much more used on the European continent than in Britain, and there is wide variation between different hospital units in Britain. As the methods are alternatives their use is largely dependent on local enthusiasm and experience, although there are a few situations where one or other is better. The ventouse is better than the forceps if urgent delivery is required and the cervix (in a first-time mother) is soft and thin and almost fully dilated. It can be used with a little less dilatation of the cervix in a multigravid patient.

A vacuum is applied to the baby's head. The scalp, which is fairly freely mobile over the underlying skull bones, slides into the cup and swells to make a knob grasped by the cup. Gentle pulling on the cup during a contraction and while the woman is pushing down aids delivery. The pull is not really on the baby's skull and doesn't harm the brain. The pull is trasnferred to the scalp's attachments at the base of the head. The knob (chignon or bun) on the baby's head disappears in a day or two. Local anaesthetic or pudendal nerve block is used but occasionally a general anaesthetic is needed.

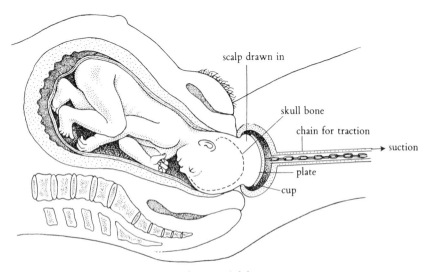

scalp drawn in

skull bone

chain for traction

suction

plate

cup

Ventouse (vacuum) delivery

Afterpains

These tend to be more of a problem for women having a second or subsequent baby. They are cramp-like pains in the lower abdomen like period pains or the pain of labour contractions. They are due to strong contraction of the muscle of the uterus sometimes associated with the passage of small blood clots. The pain can be very severe and some women have even described it as worse than labour pains. They may come on shortly after delivery in response to the injection given at delivery to make the uterus contract. The baby suckling at the breast during the first few days is the other noticeable stimulus. Paracetamol or more effectively codis cocktail repeatcd a few times will be helpful but it doesn't always relieve the pain. It should settle after a few days. (Codis cocktail contains magnesium trisilicale, peppermint water, codeine, aspirin and hot water.)

Caesarean birth (caesarean section or caesarean operation)

In Britain this is currently the method of delivery of about 7 per cent of babies. It is used when vaginal delivery would be more hazardous than caesarean section for either mother or baby. In some circumstances it is fairly easy to advise between trying for vaginal delivery or having a caesarean section, but in many cases it is very much a matter of professional judgement and expertise, based partly on generally agreed principles but partly on many years of experience of obstetrics. Now that the operation is so safe, the serious physical disadvantages to the mother are less, but it is still a major operation. It has uncommon, though well-recognized, complications and one has to consider the post-operative pain and the general reactions of the woman and her system to an abdominal operation. Much of the information we have to help us is based on statistical analysis of probabilities and can to some extent be conflicting. We try to keep you and your wife fully informed about the basis for our decisions, but inevitably you have to take much of our advice on trust.

As a caesarean is now so relatively common, you should know a little about the operation, and its place in obstetric practice, to avoid being taken unawares if it turns out to be appropriate for your wife and baby. As it is now a relatively very safe operation with few complications, it is appropriate to use it after careful evaluation of the points for and against in a particular situation, rather than as the last-ditch measure of former days.

The origins of caesarean section are lost in the mists of time, but it is so named as it was legalized during the reign of the Roman Caesars to attempt to save the baby when a woman died undelivered. The first-recorded wholly successful operation was in 1500 AD by the sow-gelder, Jacob Nufer of Siegertschaufer (Germany). It is recorded that when 13 midwives and several lithotomists (experts in cutting into the urinary bladder to remove stones), had failed to help his wife, this courageous husband took a razor and delivered a child who lived for 77 years. His wife recovered and later bore twins and four more children who were all delivered vaginally. Unlike this isolated instance, caesarean section in the past almost invariably resulted in the death of the mother from haemorrhage or infection so that destruction of the fetus was usually the preferred management if labour became obstructed.

In the mid-nineteenth century general anaesthesia was introduced, and in the latter part of that century antisepsis and asepsis were established, but it was not until relatively recently that the caesarean operation became safe.

It is worth remembering that before 1935, when the sulphonamides were introduced, there was no effective treatment for infection. This is why in 1930 there was a 15 per cent mortality among women delivered by this method after prolonged labour. Nowadays, even though the operation is sometimes necessary in women severely ill for some other reason, the overall mortality is less than one for each thousand operations.

This dramatic improvement has been due to sociological and general health improvements, but also very much to medical advances. There are now many very effective antibiotics for treating infection. Safe blood transfusion in the UK stems from the establishment of the National Blood Transfusion Service in 1941. There have been spectacular advances in the understanding of the needs of the body for fluids and their dissolved salts. The same applies to modern anaesthesia and anaesthetic techniques. Surgical methods and materials have also improved and there is continuing and progressive refinement of all aspects of practice. All these advances have enabled more ready recourse to the operation before a serious situation has been allowed to develop.

Despite all this vaginal delivery carefully controlled and managed is still greatly to be preferred when on balance it is the indicated delivery route. I should just mention that, in addition to the 'medical' considerations, a woman's preference for caesarean rather than vaginal delivery can be a factor in tipping the balance towards the former. There are obstetricians in some parts of the world who are prepared to accept a much greater contribution of patient preference in this regard.

General anaesthesia is used for most caesareans but there is a growing use of epidurals. In some places, where there is a good epidural service with competent anaesthetists, it is the most-used anaesthetic. The first caesarean section under epidural at my hospital (University College Hospital, London) was performed in 1952. I happen to know this as I by chance met the lady concerned when she was standing behind me in a queue in Regent Street last year while waiting for the January sales to open. The delivery was filmed by the BBC and was shown to her son on one of his birthdays.

Unfortunately the epidural for caesarean section needs to anaesthetize a larger area than when used for first-stage pain relief. There may not be the time necessary to get it working sufficiently well if an urgent need for caesarean arises in a woman who already has one working for relief of labour pain. A general anaesthetic is added in these circumstances. There is also the problem that an anaesthetist with the extra expertise needed may not be at hand. When the need for delivery is not so urgent, then epidurals may be extended and adapted or started afresh. It is relatively much easier to have one when caesarean is fixed for a time before labour has started (elective caesarean). This may be, for instance, when the inside dimension of the bones of the woman's pelvis are less than average and she has previously had a caesarean, or she has a baby coming breech end first.

Epidurals are not used if the woman prefers to have a general anaesthetic. If she agrees to have an epidural then there are several advantages if it is working well. You may be there with your wife, sitting at the head of the operating table separated from a view of the operation by a low screen. (Usually a junior nurse is assigned to chat to and watch you carefully, in case the occasion becomes too much and you show signs of fainting!) Being there enables you to share with your wife the excitement and joy of your baby's arrival. At the epidural birth you will be able to nurse your baby if he is well.

Your wife should not feel pain during the operation but will feel pulling and pressure and a few odd sensations. She sometimes needs an injection for nausea and may need additional pain relief if the epidural hasn't taken fully. Occasionally she will have to be given an additional general anaesthetic at which stage you will probably have to leave the operating room.

You will be surprised how quickly the baby emerges in the operation. Remember that he is likely to have some blood smeared over him from the incision in the uterus. He must at least have his nose, mouth, and throat sucked free of mucus and blood, but if he doesn't breathe and cry at once the paediatrician to hand will resuscitate him.

There are several other advantages of the epidural. Bleeding at the operation site is usually less and your wife doesn't have as much drug effect afterwards. This usually means that the two of you can have an hour or so together with your baby before your wife needs pain-relieving drugs as the epidural wears off. Sometimes the epidural is kept going longer as an alternative to these drugs, but there can be practical problems in prolonging the epidural in this way. Even without this prolongation your wife may find that her nipples are still numb when she attempts to put the baby to her breast. Disadvantages of the epidural include the occasional necessity to have to add a general anaesthetic if it isn't working well enough, or because of technical problems with the operation; occasionally your wife may wish in retrospect that she had had a general anaesthetic because she has experienced too much pain and/or anxiety; and there may be problems in passing urine postoperatively. On the whole most people who have had epidurals are delighted with them.

The main advantage of general anaesthesia, as far as the woman is concerned, is the more certain elimination of pain and anxiety and being oblivious during the surgery. Some women very reasonably prefer this and should not in any way feel obliged to have an epidural. For the obstetrician, the general anaesthetic gives somewhat better conditions for operating and less time is consumed as sometimes there can be a long wait for the epidural to give the necessary full anaesthesia. He is happier with general anaesthesia for women not temperamentally suitable for an epidural, but if he is used to working with an epidural he secondarily enjoys the enjoyment that couples can experience at the birth of their baby. Although you will not be in the operating theatre if general anaesthesia is used, the baby can be brought to you without delay.

Waking from a general anaesthetic—especially if at the same time there is considerable pain—is sometimes a nightmarish experience. The aim is to give some additional pain relief at this time but it has to be given sparingly as the anaesthetic drugs are still working. The first hours can be confusing without much recollection, but many women remember waking with the baby alongside them and they like this memory. You will probably be shocked to see how ill your wife looks as she is coming out of the anaesthetic. She may look ashen pale and be feeling sick and perhaps vomiting. You can usually sit and hold her hand and talk to her about the baby as she comes to. She will seem better as the confusion of the anaesthetic disappears.

You may be interested to know what happens in the course of the caesarean operation. The uterus lies with its outer surface against the inner surface of the front wall of the abdominal cavity with no bowel in between. This cavity is more a potential than an actual space for it contains just sufficient fluid to allow the various organs such as the uterus, ovaries, tubes, and intestines to slip on each other and on the abdominal wall as they rub in contact. During the operation air enters this space, and as it is irritating and not absorbed for a few days, it can cause quite a bit of pain in the upper abdomen, lower chest, and in the shoulders. The intestines which are disturbed by the operation contract less smoothly for a few days so that your wife gets colicky (wind) pain between the second and fourth days. These are annoying but not serious.

The object of the operation is to open the lower part of the uterus through the abdominal wall and to deliver the baby and placenta in a way which maximizes safety for mother and baby.

The skin cut is made either crosswise on the edge of the hair-line (bikini cut, Pfannenstiel incision) or up and down in the midline below the umbilicus. In general the bikini cut is most common nowadays. It has the advantages of leaving a less noticeable scar which tends to heal better. It is a stronger scar in that the different layers are cut and divided in different directions and, when resewn, reinforce each other. It causes less pain after operation and therefore allows your wife to be up and about more easily. Your wife may wonder why the abdominal wall is tender and sometimes swollen up to the level of the umbilicus. This is because after opening the skin and fat layers from side to side low down, the operator tunnels upwards in the muscle layer and spreads the wound. The opening into the abdominal cavity is made in the midline and reaches nearly to the umbilicus.

The bikini cut has some disadvantages. The little extra time it takes can be unacceptable if the baby is acutely short of oxygen, as in some cases of prolapse of the umbilical cord or acute fetal distress. If caesarean delivery needs to be repeated with subsequent pregnancies in those

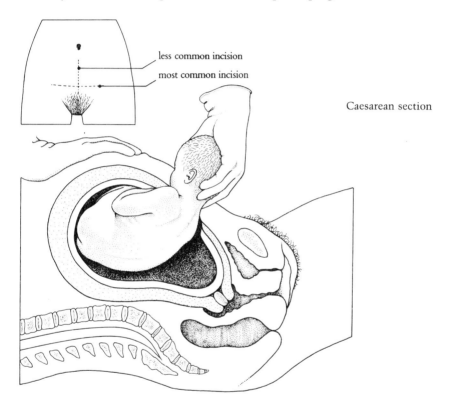

less common incision

most common incision

Caesarean section

wishing to have more than two children, the operation can become more difficult because of scarring of the abdominal wall and there being less space available for the safe delivery of the baby. Occasionally blood collects between the layers of the abdominal wall after the operation as a result of oozing, (haematoma formation). It may need draining through the incision. Finally, the wound is more likely to become infected, especially if the operation is needed after labour has been in progress for quite a long time, and for some conditions the amount of room available to operate is insufficient for safety.

The midline incision can become a less attractive scar and can very occasionally need restitching a week after operation or a hernia can develop in it over the longer term. The midline incision has the advantages of being a little quicker, giving the better exposure sometimes needed and being less liable to haematoma (accumulated blood) and infection.

The wall of the uterus is nearly always cut from side to side low down after pushing the bladder downwards from this lowest part of the uterus (the lower segment operation). The wall of the uterus is sewn in two layers and then the bladder moved back in front of the uterine scar to stop the possibility of any complication such as the intestine sticking to the scar. Not surprisingly, the bladder wall tends to be a bit bruised and this is the usual reason for pain on passing urine for one to three weeks.

Rarely we use an up-and-down incision in the uterus if the lower one cannot be made safely. In this case your wife would need to have any subsequent deliveries by caesarean as the scar in the uterus isn't reliable in labour. In closing the abdominal wall two or three layers of stitches are made beneath the skin. Even though movement can be very painful after the operation, your wife is reassured that movement and doing what she can won't interfere with healing. The skin closure is for appearance and isn't anything to do with the strength of the wound. We may use small metal clips or separate stitches or a single thread of nylon under the surface. All are removed about the fifth day for a bikini cut, and the seventh for an up and down. Your wife will be encouraged to walk the day after operation and she will find she is able with some effort and pain to stand erect from the start. If she stands straight the wound feels better.

If your wife is delivered by caesarean then her chances of successfully breast-feeding are little altered, but she will need much more help when she comes home with the baby. As I have already discussed, all women need plenty of help with looking after the household when they first return home with a new baby, but for a woman after caesarean section this is absolutely vital. She has had a major abdominal operation comparable to an abdominal hysterectomy. Most women having hysterectomy are admittedly a little older, but for them we advise a good two months' convalescence. Your wife has not only her operation to cope with but has the very onerous and demanding additional task of looking after and feeding her baby. When she leaves hospital between the eighth and twelfth days she may think herself fit for the job ahead, but all the problems of tiredness and exhaustion are likely to be multiplied. She will gradually gather strength and soon become quite fit but it may be up to six months, and sometimes longer, before she feels as fit as she would like to. It is well known that the body healing processes and reactions, as well as having a relatively fast phase, have this underlying slow long-term recovery cycle.

Your wife will be sitting out in a chair and walking around a little the day after operation, and will gradually extend her range of activities. (This is usually easier after the bikini cut.) However, she will for the first few days have difficulty in leaning over and lifting her baby. She will find it helpful to put an extra pillow under her abdomen to support it when she lies on her side. The nurses will show her comfortable positions for breast-feeding. After the first week she should gradually increase her exercises to strengthen her tummy muscles and you should encourage her with them. The time of resumption of sexual intercourse is limited by your inclination as a couple, by tiredness and general exhaustion, and by pain in the abdominal wall, but four to six weeks or so is the usual range.

While it is only reasonable that many or most couples would like to be awake and aware when their baby arrives if this is feasible, there is no evidence that the development of warm emotional bonds between parents and baby (see also p. 130) are less secure in the long run when delivery has been by caesarean section under general anaesthetic. The most damaging influence is likely to spring from unrealistic expectations. If caesarean section has not been prominently incorporated into your range of expectations for delivery, you and your wife may be bitterly disappointed and upset. It is this reaction which may set you back for a time. Try always to be realistic in your expectations and accepting of the inevitable. The decision to have a caesarean is made very much in the best interests of you all.

What are the main reasons for doing caesareans? It is very difficult to give you a list because, as I said earlier, each case needs individual consideration and careful balancing of the probable benefits of the operation against the possible risks. As the latter diminish, so it becomes reasonable to resort to caesarean rather more readily when the baby is at risk of damage. Because we can't always be sure of the risk that is present, some operations inevitably in retrospect are realized to have been unnecessary in that the baby is fitter than expected. This is a relatively small price to pay for averting the possible serious damage which could have befallen the baby.

Some of the underlying reasons for caesareans are:

1   Delay in the progress of the first stage of labour not responding to the usual measures such as rupturing of the membranes or stimulating the uterus with a syntocinon drip, when there is no clear evidence that the bony passage is too small.

2   Good evidence that the bony passage is too small, often revealed by

delay in labour, but sometimes evident before labour through assessing the size of the baby and the internal dimensions of the bony pelvis.

3   Evidence that the fetus is becoming so short of oxygen that she is running a significant risk of brain damage or even dying if labour is continued in the hope of achieving vaginal delivery.

4   Previous caesarean section, breech presentation, or twin pregnancy when there is likely to be a tight fit between the fetus and the mother's pelvis.

5   Placenta praevia (p. 171), prolapse of the cord, or a lump in the pelvis obstructing delivery (such as a low-lying fibroid).

6   Seriously undergrown baby.

7   Severe toxaemia or other medical conditions of the mother.

8   Many other uncommon conditions and situations.

After one caesarean is it necessary to have another for a subsequent birth? In general, if the indiciation for the first caesarean is no longer present and there is good evidence that there will not be a tight fit between baby and mother, then we anticipate a carefully monitored labour and vaginal delivery. If the labour does not progress well, then the attempt for vaginal delivery is abandoned in favour of repeat caesarean section.

Sometimes we can tell you at the time of the first caesarean section that a repeat operation is advised unless delivery is premature the next time. This may be on the basis of feeling the dimensions of the bones, possibly with the additional help of an X-ray in the week after delivery, or because of prior knowledge of the pelvic dimensions. Infection of the uterus during healing predisposes to repeat caesarean section. On other occasions, it will depend on how big the next baby turns out to be and a decision will be made in late pregnancy. If by the time of the next pregnancy the pelvis hasn't been assessed, then your wife may need an X-ray four to six weeks before delivery.

Sterilization in the form of a tube tie can be done very easily and quickly at the time of a caesarean section. Sometimes this is not such a good move as it seems. The couple need to have had plenty of time to consider the issues carefully. Even if the baby appears quite normal at delivery and free of defects, he can die in the next few days from unexpected or undetected defects. You may as a result want another baby in the future. As the sterilization is only *unreliably reversible* by an abdominal operation, this should not be your fall-back position. Unless you are sure you would accept such a death without wanting a reversal operation, it is better to wait and have the sterilization in a month or two when it usually means only a short stay, bringing your baby in with you. There are many other aspects of sterilization counselling but I have raised this possible disadvantage as it is so easily overlooked.

Stillbirth and neonatal death

To have a stillborn baby is an experience which I hope you and your wife will be spared, but as stillbirth is the outcome in Britain in about 0.5 per cent of births I am sure you will agree you should know something about it. It is also important to have some advance knowledge of how you and your wife would be likely to react and also to know something of the grieving or mourning process which follows.

Stillbirth is the term we use when after the 28th week of pregnancy a baby is born without showing signs of life. Usually you will know beforehand that the baby has died (intra-uterine death). At other times, though fortunately rarely nowadays, your baby can die in labour shortly before birth so that the loss is quite unanticipated.

The prevention of stillbirth is one of the most important aims of antenatal care. Most selective inductions of labour are to prevent stillbirth by delivering the fetus when there is a suspected threat to its well-being.

Prevention of fetal death during labour is a principal reason for caesarean delivery, either decided on before labour or resorted to during the course of labour.

About 50 per cent of intra-uterine deaths occur because there has been an error in the baby's development (congenital malformation). Sometimes there has been an unaccountable failure of the placenta to supply the fetus with the necessary oxygen and nourishment. The death is also more likely when the mother has toxaemia of pregnancy, high blood pressure, kidney failure, sugar diabetes, rhesus incompatibility, twins, haemorrhage from the placenta, and many other even less common problems. However, most times when one of these conditions is present the baby is born alive and well.

The underlying cause of deaths during labour may be any of the above plus the occasional case where the umbilical cord prolapses when the membranes rupture. However, we usually get warning that such a death could occur when we detect evidence that the fetus may be becoming short of oxygen (fetal distress). We respond by delivering the baby more quickly by caesarean or with forceps or the ventouse, or by making an episiotomy—which ever is appropriate.

The first inkling that all is not well in pregnancy is the loss of fetal movements. If during the last ten weeks of pregnancy your wife does not notice any movements for eight to ten hours, then we like her to report this. We then listen for the fetal heart with an ear trumpet or electronic listener (Doptone or Sonicaid) and perhaps make an electronic recording of the baby's heart rate and rhythm. Most times a false alarm is declared, but if the heart sounds are not heard we may have a final check for fetal heart movement with an ultrasound machine. If fetal death is confirmed then we tell your wife and send for you if you are not there at the time. It is not usually practicable to wait for you to arrive before telling your wife, although I agree this would be ideal. When fetal death occurs in labour or in association with a known predisposing condition, confirmation of the death is along similar lines.

Although there would be no physical harm to your wife if she were to wait for up to three weeks for labour to start spontaneously (as it almost always would) most women in this stage opt for induction now that we have fairly efficient methods for achieving it, even though it sometimes takes two or three days to effect. Some couples prefer to wait a day or two together before the induction to start to find their feet a little after the shock and initial upset of receiving the very sad news.

The procedure is to place prostglandin pessaries in the upper vagina to start the cervix opening—this may require repeated doses. When the cervix has opened a little (become more favourable) more of the stimulating substance prostglandin is passed as a solution through tubing which has been passed through the cervix into the uterus to lie between the membranes and the uterine wall. A drip of Syntocinon may be run into a vein as an additional or alternative stimulant.

Labour then proceeds much as it would normally do and your wife can have pain relief as freely as required including an epidural block.

If you can be with your wife during the labour and at delivery, you can be of great comfort and support to each other.

The reactions of you and your wife to learning of the death of your unborn baby and all your subsequent reactions are similar though somewhat coloured by your own personalities and previous life experiences. I shall therefore ask you to assume that 'you' and 'your' refer to you both.

Your initial reactions are of stunned disbelief. Sadness and weeping may come at once or be delayed for a time—perhaps until you are alone or just with your wife. Men are often trained, perhaps wrongly, to suppress emotion and grief of this kind. You should not try too hard to suppress this sadness if this is the way you feel. Expression of these feelings is an important, normal, and helpful coping mechanism. You may be concerned that you will upset your wife more by crying in her presence. On the other hand, if your wife doesn't see this genuine expression of emotion she may think you unfeeling. It will be of the greatest help to your wife if you are able to weep together. Sharing grief can be enormously calming.

You will have many questions which you should both put to your attendants, but don't be too cross when you get only the answers 'possibly' and 'perhaps', without a clear reason. The reason may be immediately apparent, at the time of delivery, as in the case of a developmental defect, or be revealed at a post-mortem later. You may subsequently need expert obstetric or genetic counselling to learn about possible recurrence with later pregnancies, and many are helped in coming to terms with their loss by bereavement counselling—discussion which helps you gradually come to terms with your loss.

Considerable research has been undertaken to find out what may help or hinder couples in this loss situation. In the past, when deaths of people of all ages were much more common, community attitudes were quite different. People were more used to and more comfortable with death. The community in general was much more supportive and able to help those bereaved to mourn their loss. Death has now become almost a taboo subject, and when a death does occur there can be a tendency to want to sweep it aside and tidy it up as soon as possible. People sometimes behave as though it hasn't happened—often because they don't know how to handle the situation. Hospital staff have tended to want to spare the parents of a stillborn baby what is regarded as unnecessary emotional pain. The baby has been spirited away as quickly as possible after delivery and an impersonal hospital-arranged burial or cremation recommended. The grave is in these circumstances unmarked and in common ground, usually unattended.

It is now realized that you need to give form and identity to your child so that you have a reality on which to fix your memory and imagination. It has been recognized that you need to create a history and a memory of your child and in the future months and years you will be grateful for this. If you have a camera, take it along with you. The hospital may produce a polaroid camera from the neonatal unit if you ask for a photo of your baby. Photographs and slides can be a great source of comfort and also enable grandparents to see their lost grandchild. Without a photograph, you will be dismayed how quickly the image of your baby disappears to leave a void in your mind. At the time you may not feel up to having a photo, but later you will be grateful. Some hospitals are starting to make a routine of taking a photograph which can be kept in your notes to be available later should you want it.

Although you should not feel obliged to do so, it can be very helpful, if after overcoming any initial revulsion, you are able to touch and hold your dead child. You have a chance to say goodbye to the child you expected. Even if there is a malformation, this can be sensitively explained to you and covered if you don't wish to see it, although usually malformations are not as bad as you imagine. Sometimes you are not ready for this right away—you need time and sensitive coaxing and explanation. You may like to both be essentially alone for a time with your baby, though it is comforting to have your midwife or attendant within easy reach.

You may think that holding a service and having a private burial is an extravagant and unnecessary way of commemorating your lost child. Giving him a name and a marked grave at which you can visit and mourn together can help more than you imagine. Photographs and a grave provide the often necessary foci for recognizing and sharing your grief.

Sometimes a couple is unable or prevented from sharing grief. A man who has been shut off from the reality of stillbirth or has suppressed his emotions and perhaps brooded quietly is less able to share with his wife in this way. He is more likely to suffer considerable depression. Maritial discord or even marital breakdown can follow.

After the delivery your wife will usually spend one to three days in hospital, depending on your home circumstances, family support, and your wife's health. Many people are lucky in having friends who will visit and genuinely sympathize and listen after the return home. You may well find that you, and especially your wife, will want to go over and describe your ordeal to everyone who will listen. You may go through a phase of being angry that this tragedy should have befallen you, and ask, 'Why us?' Sometimes you will blame those who have been looking after your wife—sometimes unreasonably, but sometimes sadly with some justification.

You and your wife will go back over previous events relating to the pregnancy trying to find a reason for the loss and wonder what you could have done to prevent it. Sometimes your wife will think of a time when she hadn't wanted the pregnancy, had thought of having an abortion or even tried to bring it on with a home remedy such as jumping down the stairs: you may think of sexual intercourse as a cause or of having smoked or indulged in alcohol or unusual foods—or perhaps your wife shouldn't have worked so hard or rested more. It is easy to see how you can be overtaken by guilt or remorse. As it happens, it is extremely unlikely that any of the above are remotely connected, but you should discuss and ask questions about anything which bothers you.

After the first week or two, you may find yourselves becoming depressed and apathetic, and wondering whether you will ever be able to have a child.

You may well find some friends avoiding you rather than having to talk to you about the stillbirth. It is often that they don't know what to say and are afraid of upsetting you. All that is needed is for them to say how sorry they are that you have lost your baby. Others will try to carry on a conversation as though nothing has happened. You can sometimes make it easier for them by being willing to talk even though you find it painful. Some will try to help you and your wife by trying to jolly you out of your moods, or by suggesting after a couple of months that it's time to forget about it and look to the future. This is no doubt well-meant but premature. You can never forget, although you do come to accept it as part of your life and you do adjust to it and to the future. The anniversary of the baby's loss can be a difficult time when grief tends to return.

Should you try for another baby as soon as possible to help you forget? This is seldom wise. Much better to give yourselves about 9–12 months to come to terms and to grieve adequately. Grief cut short by another pregnancy too soon can sometimes return and alter the relationship with the next child—he becomes a replacement rather than a separate individual.

A pamphlet *The loss of your baby* is produced by the Health Education Council. The Stillbirth and Neonatal Death Society (SANDS), 37 Christchurch Hill, London, NW3. Tel: 01 794 4601) is a self-help organization with branches in many parts of Britain. You will find a group among whom you can ask questions and compare grievances without being a bore. Most of these people have made a good recovery and are able to direct their compassion and skills towards helping others. Even if you don't need their help you might like to join and help others in the future.

Neonatal death

In the strict statistical sense this is death in the first week of life, but for your purposes means death of your child within the first few weeks. Everything I've said about stillbirth applies equally well here. Many babies lost at this time are handicapped because of malformation or a problem around the time of birth, while others are immature. You will usually have some warning because your baby is not well but no matter how much it has been anticipated, the full realization of your loss comes only when your baby dies and then you have a long period of distress and mourning ahead of you.

Depression after the birth

Depression is an illness which may affect any one at any time, but women are particularly prone to it in the first few months and up to a year after giving birth. During this period between 3 and 10 per cent of women need the help of a doctor because of depression. You need to be aware of the possibility so that you detect the early manifestations. You may notice in your wife a reduction in the usual mood swings—a lowering or flattening of mood. There may be sleep disturbance, a tendency to weep easily, inability to concentrate, and a poor memory. There may be over-anxiety and panic attacks. It is easy to put all this down to over-tiredness and to worry about the baby. Tiredness does cause the same symptoms, but with depression sleep does not bring the expected improvement. Depression, of course, does bring difficulty in coping with the baby. Your wife will lack the emotional elasticity needed to cope with her baby and also with your evotional and sexual needs. It may seem to you that you have lost contact as a couple. The changes usually develop slowly, so that you are often the last person to realise that her personality changes may be due to this treatable illness.

If you are in doubt, you can discuss your concern with a close friend and certainly with your family doctor or health visitor. Your wife may not agree to see your doctor, but then you can usually arrange for him to make a house call or for your wife to make a visit to the doctor's surgery for help with the baby.

Prevention is quite clearly the best approach to this problem. Even though we are not always sure why this illness develops, your support and understanding through pregnancy and afterwards can be most effective in alleviating the stresses of this changed life situation. There is some evidence that some women have a background which makes them more likely to develop depression at this time. There are many situations such as having come from a home where there has been excessive quarrelling, and particularly when a girl has not had a good relationship with her father; or when she has been deprived of one or other parent from a relatively early age.

The mere fact that you are troubling to read this book is an indication of your concern. Do not hesitate to seek help at any stage if you sense the need. Your supportive role, which is discussed in many sections of this book, is the basis of your contribution to prevention. You can do much to foster a general sense of relaxation and confidence, rather than inadvertently reinforcing any feeling of incompetence that your wife may be developing. However, despite your best efforts, this problem can arise quite unexpectedly, and one cannot assume that anyone is immune.

Glossary

alpha-fetoprotein (AFP) A protein formed in the fetus and present in small amounts in the amniotic fluid and in the mother's blood. In certain congenital abnormalities the level of AFP is raised.

amniocentesis The withdrawal of a small amount of amniotic fluid.

amniotic fluid The 'waters' surrounding and protecting the fetus.

antenatal During pregnancy before the baby has delivered, and usually excluding labour.

cell The basic building block of the body. Cells are combined by the intercellular substance containing binding material to make the tissues of the body. Each cell has a central compartment called the nucleus. Each human cell nucleus contains 46 chromosomes. In each individual these sets of chromosomes, which determine form, structure, and function, are identical in every cell nucleus throughout the body.

cervix The lower narrower part (the neck) of the uterus. It contains the narrow passage from the vagina to the upper uterus.

chromosome The potential for expression of individual characteristics acquired from the parents is carried in the genes which are aggregated to form the 23 pairs of thread-like chromosomes. Each chromosome has characteristic form and staining and can be identified and studied when dividing (multiplying) cells are specially prepared to display the chromosomes.

diabetes This is a group of disorders in which large amounts of urine are passed. Diabetes mellitus (sugar diabetes) is the one which concerns us. There is reduced capacity to produce the hormone insulin which controls the levels of glucose sugar in the blood and the use of glucose by the cells for energy production. Blood glucose levels rise and the glucose lost in the urine brings water with it to increase the volume. Treatment is to give insulin injections and to control the sugar content of the diet.

embryo The future baby in the first ten weeks of development during which time all the main organs are formed.

fibroid A benign localized growth of the wall of the uterus.

fore-waters The liquor contained within the membranes below the presenting part of the baby. Approaching from the vagina and through the cervix, the membranes and their contained fore-waters are met before the first part of the fetus. These are drained to induce labour or to accelerate a desultory labour.

high forceps The method used in former times to deliver a baby when in the second stage of labour the head failed to engage in the pelvis. A difficult and dangerous procedure now replaced by the much safer procedure of caesarean section.

hind-waters   The portion of the liquor above the presenting part. One theory to explain cases in which a leak of what is presumed to be liquor stops and the pregnancy continues is that the membranes have developed a hole high up. This allows escape of hind-water but the hole in the membranes heals because the membranes are against the wall of the uterus.

hyosine   A drug which was used for many years earlier in this century in combination with morphine to produce the 'twilight sleep' during the first stage of labour. Unfortunately large doeses of both drugs were often needed and some women became excited and confused and had to be restrained. They were sometimes left with strange disturbed mental states and poorly formulated fears. It is no longer used.

jaundice of the newborn   The accumulation of the yellow pigment bilirubin usually because the liver is not yet mature enough to eliminate it, sometimes because of rhesus or other problems.

laparoscopy   Inspection of the abdomen with a telescope after filling it with gas after inducing an anaesthetic.

L.O.A. (L.O.L., L.O.P., R.O.A., R.O.L., R.O.P.)   Left-occipito-anterior. These are ways of describing where the back (occiput) of the baby's head is pointing. L.O.A. means pointing to the front on the left but it can be lateral (to the side) or posterior (more backwards). These are called the positions of the baby and are of some interest in following the progress of labour. Labour may proceed differently with the different positions.

meconium   The content of the bowel in fetal life and in the first few days after birth. It is dark green because of the high bile content. Passage of meconium into the liquor in labour suggests that the fetus could be short of oxygen. It is revealed only after breaking the waters.

morphine   The principal active ingredient of opium which is extracted from the juice of the opium poppy. The pain-relieving properties of this juice have been known from pre-historic times.

moulding   The changing of the shape of the fetus's head during labour.

multigravida   Used loosely to refer to a woman who has had at least one baby previously.

ovum   The egg or germ cell liberated from a woman's ovaries each month. Fertilization in the outer part of a Fallopian tube by incorporation of a male sperm sets off the process of cell division which results in the development and birth of a baby. The ovum is the size of a small full stop.

oxytocin   A naturally occurring hormone which affects the contractions of the uterus. It has been manufactured artificially and is used to induce or speed up contractions of the uterus.

Pavlov   His study of conditioned reflexes in dogs influenced behaviourism. A conditioned reflex is a response which becomes dependent on an event in the environment. In psychoprophylaxis for labour, the woman is trained to respond in a particular way when she feels a uterine contraction starting.

perineum   As commonly used in obstetrics, this is the ill-defined area on either side between the back end of the vulva and the anus. Under the skin, the muscles of the area come together from side to side. The vagina lies to the front of the muscle and the lower bowel (anal canal) behind. The perineum becomes greatly distended during birth. An episiotomy is made when a serious tear seems likely.

physiotherapist A person trained to use physical and natural methods for relief and cure. Some physiotherapists have had extra training in preparation for labour work and getting back to normal afterwards.

pituitary gland A small organ associated with the base of the brain. It produces hormones which either directly, or through other glands such as the thyroid or ovaries, control many of the body's functions. It produces the oxytocin which is involved in the contraction of the uterus in labour, and also causes the let-down of milk for breast-feeding.

placenta (afterbirth) The organ of exchange between the mother and fetus. The baby gains oxygen and food and excretes waste-products. It is on the wall of the uterus and the baby is attached to it by the umbilical cord.

presentation, presenting part The lowermost portion of the fetus—the portion which 'presents' to the birth canal or will come first during labour. Usually the head presents (cephalic presentation). The head is usually well tucked up so that the top of its back (vertex) presents, but if the chin is less tucked down it can be a face or brow (forehead) presentation. If the other end (breech) presents there are likewise several variations—'footling' if it is the feet, 'frank' if it is the bottom, and 'complete' if it is both feet and bottom. Each variety of presentation has its own importance and implications for delivery management.

premature This is a difficult term as it has been defined in terms of duration of pregnancy (less than 37 completed weeks) or baby size (2500 grams (5½ lbs) or less). Some small babies of this size have completed more than 37 weeks in the uterus but have grown slowly. They are better called 'small for dates' (or 'dysmature') and those born early are 'pre-term'. A baby growing slowly and born early is both small for dates and pre-term. Premature labour is one starting before 37 completed weeks.

primagravida Strictly a woman who is pregnant for the first time. However loosely used to refer to a woman having her first baby and ignoring previous miscarriages etc.

respiratory distress syndrome The condition of a newborn infant in which the lungs are imperfectly expanded. It is a problem of immature babies and requires special paediatric care which may include artificial respiration for a few days.

uterus (womb) The organ designed to nourish and protect the developing fetus until he is mature enough to live outside. It is a major force in expelling him at delivery.

Index